暨南大学本科教材资助项目

分析化学实验

Analytical Chemistry Lab

谢新媛　李风煜　主编

暨南大学出版社
JINAN UNIVERSITY PRESS

中国·广州

图书在版编目（CIP）数据

分析化学实验 = Analytical Chemistry Lab：汉英对照 / 谢新媛，李风煜主编. -- 广州 ：暨南大学出版社，2025. 8. -- ISBN 978-7-5668-4155-1

Ⅰ. O652.1

中国国家版本馆 CIP 数据核字第 2025H1N361 号

分析化学实验

FENXI HUAXUE SHIYAN

主　编：谢新媛　李风煜

出 版 人：阳　翼
策划编辑：曾鑫华
责任编辑：黄亦秋
责任校对：王燕丽
责任印制：周一丹　郑玉婷

出版发行：暨南大学出版社　（511434）
电　　话：总编室（8620）31105261
　　　　　营销部（8620）37331682　37331689
传　　真：(8620) 31105289（办公室）　37331684（营销部）
网　　址：http://www.jnupress.com
排　　版：广州市新晨文化发展有限公司
印　　刷：广东诚粤印务有限公司
开　　本：787mm×1092mm　1/16
印　　张：13.25
字　　数：295 千
版　　次：2025 年 8 月第 1 版
印　　次：2025 年 8 月第 1 次
定　　价：49.80 元

（暨大版图书如有印装质量问题，请与出版社总编室联系调换）

前　言

本书的编写基于暨南大学全英文分析化学实验课程的教学实践及自编全英文讲义。本书采用中英双语并行编排模式，确保中文实验内容与其对应的英文表述一一对应，适用于化学、化工、药学、食品科学、环境科学和生物学等专业的分析化学实验教学。

本书实验分为三类：基础性实验、设计性实验和文献（创新）实验。基础性实验注重基础知识的传授，强调实验基本操作和技能的训练，涵盖常用的定量化学分析方法，包括酸碱滴定、配位滴定、氧化还原滴定等滴定分析方法及应用，以及 pH计和分光光度计等常用仪器的原理与使用方法。本书还包含以下内容：分析化学实验的基本知识、实验基本操作、实验数据的采集与处理等。特别地，本书引入了来自实际案例和科研工作中的设计性实验和文献（创新）实验，旨在培养学生的创新实践能力，提升其发现问题、分析问题和解决问题的能力。

为拓展教学维度并提升学习便捷性，本书拓展资源可通过网站（https://jnupress.com/download/index）获取，包括沉淀重量分析仪器及使用方法、利用 WPS 电子表格处理分析化学数据、仪器基本操作视频、部分实验的教学辅助课件，以及分析化学实验常用数据。

本书的出版得到暨南大学化学与材料学院的大力支持。学院实验中心岳攀老师参与了本书第二章的编写，并制作了相关的基本操作视频。本书撰写过程中得到了分析化学教研室白燕教授的指导和帮助，同时参考了多本相关教材和文献资料，在此一并致以诚挚的感谢！

由于编者水平有限，书中难免存在疏漏和不足之处，恳请读者批评指正。

编　者

2025 年 5 月

Preface

This textbook is developed based on the teaching practices of the full-English Analytical Chemistry Experiment course at Jinan University and self-compiled full-English lecture notes. It adopts a bilingual (Chinese and English) parallel arrangement, ensuring that the Chinese experimental content corresponds accurately to its English counterpart, and is suitable for teaching analytical chemistry experiments in various majors, including chemistry, chemical engineering, pharmacy, food science, environmental science, and biology.

The experiments in this book are categorized into three types: foundational experiments, design-oriented experiments, and literature-based (innovative) experiments. Foundational experiments emphasize the transmission of basic knowledge and focus on training fundamental operational skills and techniques. These include common quantitative chemical analysis methods such as acid-base titration, complexometric titration, redox titration, and other titrimetric analysis methods along with their applications. Additionally, the textbook covers the principles and usage of commonly employed instruments like pH meters and spectrophotometers. The textbook also includes essential topics such as basic knowledge of analytical chemistry experiments, fundamental operations, data collection and processing methodologies. Notably, it incorporates design-oriented experiments and literature-based experiments derived from real-world cases and research work, aiming to foster students' innovative practical abilities and enhance their problem-solving capabilities.

To broaden the teaching scope and improve learning convenience, the extended content of this book can be accessed via the website (https://jnupress.com/download/index), including apparatus for precipitation gravimetry, data processing for analytical chemistry using WPS spreadsheets, basic operation videos of apparatus, selected teaching auxiliary materials for certain experiments, and common data in analytical chemistry.

The publication of this textbook has been significantly supported by the College of Chemistry and Materials at Jinan University. Yue Pan from the Experimental Center contributed to the compilation of Chapter 2 and produced relevant basic operation videos. Throughout the writing process, Professor Bai Yan from the Analytical Chemistry Teaching and Research Office provided valuable guidance and assistance. Furthermore, multiple textbooks and literature references were consulted during the preparation of this material. We extend our heartfelt gratitude to all contributors.

Despite the best efforts of the compilers, there may still be omissions or areas for improvement in this textbook. We sincerely welcome feedback and corrections from readers to enhance its quality.

Editor
May 2025

目　录

CONTENTS

第一章　分析化学实验的要求及基础知识

Chapter 1　Requirements and Basic Knowledge of Analytical Chemistry Experiments

第一节　分析化学实验的目的和要求

分析化学实验的目的是：巩固和加深学生对分析化学基本概念和基本理论的理解；使学生正确熟练地掌握常见分析方法的基本操作和技能，学会正确合理地选择实验条件和实验仪器，善于观察实验现象和进行实验记录，正确处理数据和表达实验结果；培养学生良好的实验习惯、实事求是的科学态度和严谨细致的工作作风，以及独立思考、分析问题、解决问题的能力；使学生逐步地掌握科学研究的技能和方法，为学习后续课程和将来工作奠定良好的实践基础。

为了达到上述目的，在学习过程中应注意做到如下几点：

1. 认真预习

每次实验前必须明确实验目的和要求，理解分析方法和分析仪器工作的基本原理，熟悉实验内容、操作程序及注意事项，查好有关数据，列出数据记录表格，完成预习报告。对于设计性实验，要根据实验提示及要求，查阅有关手册、参考书和文献，设计出自己的实验方案，经小组讨论及指导教师审阅后，方可进入实验室。

2. 仔细实验，如实记录，积极思考

实验过程中，要认真地学习有关分析方法的基本操作技术，在教师的指导下正确使用仪器，严格按照规范进行操作。细心观察实验现象，及时将实验条件和现象以及分析测试的原始数据记录于实验记录本上，不得随意涂改；同时要勤于思考，善于发现、分析和解决问题，培养良好的实验习惯和科学作风。

3. 严格遵守实验室规则，注意安全

保持实验室内安静、整洁。实验台面保持清洁，仪器与试剂按照规定摆放，做到整齐有序。爱护实验仪器设备，实验中如发现仪器工作不正常，应及时报告教师处理。实验中要注意节约。安全使用电、煤气和有毒或腐蚀性的试剂。每次实验结束后，应将所用的试剂及仪器放回原位，清洗好用过的器皿，整理好实验室。

Section 1   Course Objectives and Requirements

The purpose of this course is to help students understand the fundamental concepts and theories of analytical chemistry. It aims to enable students to correctly and skillfully master the basic operation and skills of common analytical methods, learn to correctly and reasonably select experimental conditions and instruments, be adept at observing and recording experimental phenomena, handle data processing and presentation of experimental results correctly. It also aims to foster students' good experimental habits, a scientific attitude of seeking truth from facts and a rigorous and meticulous work style, as well as the ability to think independently, analyze and solve problems, thereby enabling students to gradually master the skills and methods of scientific research and lay a sound practical foundation for subsequent courses and future work.

To achieve the aforementioned objectives, the following points should be noted during the learning process:

1. Thorough preview

Before each experiment session, it is essential to clarify the purpose and requirements, understand the fundamental principles of the analytical method and the operation of the analytical instrument, be familiar with the content of the experiment, the operation procedures and precautions, look up the relevant data, draft the data recording table, and finish the pre-lab report. For design-oriented experiments, it is necessary to consult the relevant manuals, reference books and literature according to the experimental tips and requirements, and design one's own experimental scheme. Only after group discussion and review by the instructor can one do the experiment.

2. Conduct experiments carefully, record truthfully, and think actively

During the experimentation process, it is necessary to learn the fundamental operation techniques of the relevant analytical methods earnestly and use the instruments correctly under the guidance of the instructor. The operations should be carried out strictly in accordance with the norms. Observe the experimental phenomena carefully and record the experimental conditions, phenomena, and the original data on the experimental notebook in a timely manner without making any random alterations. At the same time, be diligent in thinking, be adept at discovering, analyzing, and solving problems, and cultivate good experimental habits and a scientific work style.

4. 认真写好实验报告

根据实验记录进行认真整理、分析、归纳、计算，并及时写好实验报告。实验报告一般包括实验名称、实验日期、实验目的、实验原理、主要仪器与试剂、实验步骤、实验数据（或图谱）及其分析处理、实验结果和讨论。实验报告应简明扼要，图表清晰。

3. Strictly adhere to laboratory rules and be mindful of safety

Maintain a quiet and tidy environment in the laboratory. Keep the laboratory bench clean, and arrange the instruments and reagents neatly and orderly as stipulated. Cherish the laboratory equipment. If any abnormal operation of the instrument is detected during the experiment, report it to the instructor promptly for handling. Be thrifty during the experiment. Use electricity, gas, and toxic or corrosive reagents safely. After each experiment, restore the used reagents and instruments to their original positions, clean the used vessels, and tidy up the laboratory.

4. Write the lab report earnestly

Carefully collate, analyze, summarize and calculate according to the experimental records, and write the lab report in time. The lab report generally includes name of the experiment, date, objectives, principle, main instruments and reagents, procedure, experimental data (or diagrams) and one's analysis and processing, experimental results and discussion. The lab report should be concise and to the point, with clear charts and tables.

第二节　　实验数据的记录和实验报告

1. 实验数据的记录

学生要准备专门的实验记录本，标上页数，不得撕去任何一页。绝不允许将数据记在单页纸上、小纸片上，或随意记在其他地方。实验过程中的各种测量数据及有关现象，应及时、准确而清楚地记录下来。记录实验数据时，要本着严谨的科学态度，实事求是，切忌夹杂主观因素，决不能伪造或随意拼凑数据。

实验过程中涉及的各种特殊仪器的型号和标准溶液浓度等，也应及时准确记录下来。

记录实验数据时，应注意其有效数字的位数。用分析天平称量时，要求记录至 0.000 1 g。滴定管及吸量管的读数，应记录至 0.01 mL。用分光光度计测量溶液的吸光度时，如吸光度在 0.6 以下，应记录至 0.001 的读数；如在 0.6 以上，则记录至 0.01 的读数。

实验中的每一个数据都是测量结果，所以重复测量时，即使数据完全相同，也应记录下来。在实验过程中，发现数据算错、测错或读错而需要改动时，可将数据用一横线画去，并在其上方写上正确的数字。

在定量分析中，一般要求平行测定 3~5 次，通常平行测定 3 次。分析结果的精密度通常用相对平均偏差（relative average deviation，RAD）表示。

三次结果的算术平均值为：$\bar{x} = \dfrac{x_1 + x_2 + x_3}{3}$

平均偏差为：$\bar{d} = \dfrac{\left| x_1 - \bar{x} \right| + \left| x_2 - \bar{x} \right| + \left| x_3 - \bar{x} \right|}{3}$

相对平均偏差为：$RAD = \dfrac{\bar{d}}{\bar{x}} \times 100\%$

Section 2   Recording of Experimental Data and Lab Report

1.   Recording of experimental data

Students should prepare a laboratory notebook with numbered pages, and do not tear out any page. Never write down data on a single page, small pieces of paper, or anywhere else. All kinds of measured data and related phenomena during the experiment should be recorded promptly, accurately, and clearly. When recording experimental data, one should adopt a rigorous scientific attitude and be factual and realistic. Subjective factors must not be intermixed, and data must never be fabricated or pieced together randomly.

Models of various instruments involved in the experiment and concentrations of standard solutions should also be recorded accurately and promptly.

Pay attention to the digits of significant figures in the recording of experimental data. Round up to the nearest ten thousandths when weighing with an analytical balance. Read burets and pipets to the nearest 0.01 mL. When measuring the absorbance of a solution with a spectrophotometer, if the absorbance is below 0.6, a reading to the nearest 0.001 should be recorded; if the absorbance is above 0.6, it is required to record a reading to the nearest 0.01.

Each data point in the experiment is a measurement result, so when replicate measurements are made, even if the data are exactly the same, they should be recorded. In the process of the experiment, if it is found that the data is wrongly calculated, measured or read and needs to be modified, the wrong data can be crossed out with a line and the correct number should be written above it.

In quantitative analysis, it is generally required to measure 3–5 replicates, typically 3 replicates. The precision of the analytical results is usually expressed by the relative average deviation (RAD).

The arithmetic mean of the three results is: $\bar{x} = \dfrac{x_1 + x_2 + x_3}{3}$

The average deviation is: $\bar{d} = \dfrac{\left| x_1 - \bar{x} \right| + \left| x_2 - \bar{x} \right| + \left| x_3 - \bar{x} \right|}{3}$

The relative average deviation is: $RAD = \dfrac{\bar{d}}{\bar{x}} \times 100\%$

2. 实验报告

实验完毕，应根据预习和实验中的现象及数据记录等及时、认真地写出实验报告。定量分析实验报告一般包括以下内容：

（1）实验编号及实验名称。

（2）实验目的：交代实验要取得的效果。

（3）实验原理：简要地用文字和化学反应式说明实验原理。例如，对于滴定分析，通常应有标定和滴定的反应方程式、基准物质和指示剂的选择、标定和滴定的计算公式等。对于特殊实验装置，应画出实验装置图。

（4）实验所用的仪器和试剂：列出实验中所使用的主要仪器和试剂。除非特别注明，本教材所用化学试剂均为分析纯试剂，实验用水均为蒸馏水。

（5）实验步骤：应简明扼要地写出实验步骤。

（6）实验记录与数据处理：应用文字、表格、图形将数据表示出来。根据实验要求及计算公式计算出分析结果并进行有关数据和误差处理，尽可能地使记录表格化。

（7）结果与讨论：总结实验的结果，说明所有实验目的是否都已完成。可围绕以下问题进行讨论：实验结果的精密度和准确度如何？哪些错误可能影响数据？从本实验中学到了什么新技术、仪器、化学过程等？为什么这个实验很重要？此外，还应包括任何其他与实验相关的重要问题。讨论时注意要有理有据，用数据和理论支撑论点；语言客观专业，使用科学词汇，句子不要以"我想""我相信""我估计"或任何类似的表达开头，引用的参考文献要标注。

上述各项内容的简繁取舍，应根据各个实验的具体情况而定，以清楚、简练、整齐为原则。实验报告中的有些内容，如原理、表格、计算公式等，要求在实验预习时准备好，其他内容则可在实验过程中以及实验完成后填写、计算和撰写。

2. Lab Report

After the experiment, write the lab report promptly and earnestly according to the preview and the phenomena and data recorded in the experiment. The lab report of quantitative analysis generally includes the following contents:

(1) Experiment No. and name of experiment.

(2) Objectives: State the expected results of the experiment.

(3) Principle: Explain the experimental principle briefly with words and chemical reactions. For example, for titrimetric analysis, usually there should be standardization and titration reaction equations, selection of primary standards and indicators, calculation formulas for standardization and titration, etc. For special experimental setups, a diagram of the apparatus should be drawn.

(4) Apparatus and chemicals: List the main apparatus and chemicals to be utilized in the experiment. Unless otherwise noted, all the chemical reagents used in this textbook are analytical reagent (AR), and the water for the experiment is distilled water (DI water).

(5) Procedure: The procedure of the experiment should be listed concisely and succinctly.

(6) Data recording and processing: Data should be presented using text, tables, and graphs. According to the experimental requirements and calculation formulas, the analytical results should be computed, and relevant data and errors should be analyzed and processed. Records should be maintained in a tabular format as much as possible.

(7) Results and discussion: Summarize the results of the experiment and indicate whether all experimental objectives have been accomplished. The data and results should be discussed to answer the following questions: How precise and accurate are the experimental results? What errors could have affected the data? What new technique (s), instrumentation, chemical processes, etc. can be learnt from the experiment? Why is this experiment important? It should also imclude any other important issues relevant to the experiment. Make statements and then back them up using data and information. Be professional in your grammar and use scientific vocabulary. No sentence should start with "I think," "I believe," "I assume" or anything else like that. Cite any outside sources of information that you use.

The selection of the above contents should be based on the specific situation of each experiment. The final lab report should be clear, concise and neat. Some contents of the lab report, such as principles, tables, calculation formulas, etc., should be prepared during the experiment preview, while others can be filled in, calculated and written during and after the experiment.

第三节　实验室安全知识

（1）在分析化学实验室中，经常使用的设备、玻璃仪器、电器及化学药品等都潜伏着危险，如果不遵守操作规则，不仅会导致实验失败，而且可能会造成事故，如割伤、触电、中毒、着火及爆炸等。为确保人身安全和实验室、仪器设备的安全，以及避免环境污染，学生必须认真学习实验教材中有关实验安全的知识，养成安全实验的良好习惯，认真做好实验前的预习，注意听从教师的指导，在实验过程中严格执行操作规范，以保证实验的正常进行。

（2）禁止将食物和饮料带进实验室，禁止吸烟，实验中注意不用手摸脸、眼等部位。一切化学药品严禁入口，实验完毕后必须洗手。水、电、煤气等使用完毕后，应立即关闭。离开实验室时，应仔细检查水、电、煤气、门、窗是否均已关好。

（3）使用电器设备时，应特别小心，切不可用湿润的手去开启电闸和电器开关。凡是漏电的仪器不要使用，以免触电。

（4）使用浓酸、浓碱时应特别小心，防止溅出。使用浓HNO_3、HCl、H_2SO_4、$HClO_4$、$NH_3 \cdot H_2O$时，均须在通风橱中进行。若不慎溅在实验台或地面上，必须及时用湿抹布擦洗干净；若不小心将酸或碱溅到皮肤或眼内，应立即用大量水冲洗，然后用$50 \, g \cdot L^{-1}$ $NaHCO_3$溶液（酸腐蚀时使用）或$50 \, g \cdot L^{-1}$ H_3BO_3溶液（碱腐蚀时使用）冲洗，最后用水冲洗。

（5）使用汞、汞盐、砷化物、氰化物等剧毒品时，实行登记制度，取用时要特别小心，切勿泼洒在实验台面和地面上，用过的废物、废液切不可乱扔，应分别回收，集中处理。氰化物不能接触酸，否则会产生剧毒的HCN。氰化物废液应倒入碱性亚铁盐溶液中，使其转化为亚铁氰化物盐类，然后作为废液处理，严禁直接倒入废液缸或下水道内。

（6）安全使用水、电，严防火灾。一旦发生火灾，要保持镇静，先立即切断电源或燃气源，再采取具有针对性的灭火措施。一般的小火用湿布、防火布或沙子覆盖燃烧物

Section 3   Laboratory Safety

(1) In analytical chemistry lab, frequently used equipment, glassware, electrical appliances and chemicals are potentially dangerous. Non-compliance with operational protocols can lead to experimental failure, and may result in accidents such as cuts, electric shocks, poisoning, fires, or explosions. To ensure personal safety, protect laboratory facilities and instruments, and prevent environmental pollution, students must diligently study the safety guidelines provided in the textbook, cultivate safe experimental practices, thoroughly prepare before experiments, attentively follow the instructor's guidance, and strictly adhere to operational procedures during the experiment to ensure its smooth progression.

(2) It is prohibited to bring food and beverages into the laboratory, and smoking is forbidden. During the experiment, be cautious not to touch areas such as the face and eyes with your hands. All chemicals are strictly forbidden from being ingested. Hands must be washed upon completion of the experiment. Once the use of water, electricity, and gas is finished, they should be immediately turned off. Before leaving the laboratory, carefully inspect to ensure that water, electricity, and gas are turned off, as well as doors and windows are all properly closed.

(3) Special care should be taken when using electrical equipment. Do not open switches and electrical appliances with wet hands. Do not use any instrument with eletrical leakage in case of electric shock.

(4) Extra care should be taken when handling strong acids and bases, preventing from spilling out. Concentrated HNO_3, HCl, H_2SO_4, $HClO_4$ and $NH_3 \cdot H_2O$ must be handled in a fume hood. In the event of accidental spillage on the laboratory bench or floor, prompt cleaning with a damp cloth is mandatory. If an acid or base is accidentally splashed onto the skin or into the eyes, rinse immediately with copious amounts of water, then with $50 \text{ g} \cdot \text{L}^{-1}$ $NaHCO_3$ solution (for acid exposure) or $50 \text{ g} \cdot \text{L}^{-1}$ H_3BO_3 solution (for alkali exposure), and finally with water.

(5) When handling highly toxic substances such as mercury, mercury salts, arsenides, and cyanides, a registration system shall be implemented. Special caution is necessary during their retrieval to avoid spillage on the laboratory bench or floor. The used waste and waste liquid must not be randomly discarded but should be separately collected and centralized for treatment. Cyanides must not come into contact with acids, otherwise, highly toxic HCN will be generated. Cyanide waste liquid should be poured into an alkaline ferrous salt solution to convert it into

至熄灭。不溶于水的有机溶剂以及金属钠等能与水起反应的物质一旦着火，绝不能用水去浇，应用沙子覆盖或用二氧化碳灭火器灭火。如电器起火，不可用水冲，应当用四氯化碳灭火器灭火。如情况严重应立即报火警。

（7）在实验过程中不慎发生烫伤时，一般用浓的（90%~95%）酒精消毒后，在烫伤部位涂上苦味酸溶液或烫伤软膏，严重者应立即送医院治疗。

（8）应保持实验室内整齐、干净。不能将毛刷、抹布扔在水槽中。禁止将固体物、玻璃碎片等扔入水槽中，以免造成下水道堵塞。此类物质以及废纸、废屑应放入废纸箱或放置于实验室规定存放的地方。废酸、废碱应小心倒入废液缸，切勿倒入水槽内，以免腐蚀下水管。

ferrocyanide salts, and then undergo waste liquid treatment. It is strictly prohibited to directly pour it into the waste liquid tank or sewer.

(6) Employ water, electricity, and gas safely and be vigilant against fires. Once a fire occurs, remain composed. Immediately disconnect the power or gas source first and then adopt targeted fire-extinguishing measures. Minor fires can typically be extinguished by covering the burning objects with a wet cloth, fireproof fabric, or sand. Organic solvents insoluble in water and substances that react with water, such as sodium metal, must never be doused with water if they catch fire. Instead, cover them with sand or use a carbon dioxide fire extinguisher. In the event of an electrical appliance fire, water should not be used. A carbon tetrachloride fire extinguisher should be employed. If the circumstances are severe, an immediate alarm should be raised.

(7) In the event of accidental scalding during an experiment, after disinfection with concentrated (90%–95%) alcohol, apply picric acid solution or burn ointment on the burned area. Severe cases should be promptly referred to a hospital for treatment.

(8) The laboratory should be maintained in an orderly and clean condition. Brushes and rags must not be thrown into the sink. Solid substances, glass fragments, etc. are prohibited from being thrown into the sink to prevent clogging of the sewer. Such materials, along with waste paper and debris, should be placed in the waste container or the designated storage area as stipulated by the laboratory. Waste acids and alkalis should be carefully poured into the waste liquid cylinder and must not be poured into the sink to avoid corrosion of the sewer pipes.

第二章　分析化学仪器及其基本操作

Chapter 2　Apparatus and Unit Operations of Analytical Chemistry

第一节　分析天平

分析天平是实验中进行准确称量时最重要的仪器，它可以分为机械类和电子类。机械类分析天平可细分为普通分析天平、空气阻尼天平、半自动电光天平、全自动电光天平和单托盘天平等。这些天平均利用杠杆原理进行工作，只是在结构和使用方法上有所不同。此类天平的最大优点是结构直观，缺点则是零件复杂，操作要求高且费时。

近些年来，基于杠杆原理的机械类分析天平已逐渐被淘汰，取而代之的是基于电磁力平衡原理的电子分析天平。电子分析天平有即时称量、不需砝码、达到平衡快、直显读数、性能稳定、操作简单等特点。此外，电子分析天平还具有自动校正、自动去皮、超载显示、故障报警、信号输出及数据处理等功能。

一般电子分析天平的分度值为0.1 mg，即可称出0.1 mg质量或分辨出0.1 mg的差别。半微量电子分析天平的分度值为0.01 mg，微量电子分析天平的分度值为0.001 mg，超微量电子分析天平的分度值为0.000 1 mg。电子分析天平的最大载荷量一般为100~200 g。

一、电子分析天平的原理

电子分析天平是利用电磁力与被测物体的重力相平衡的原理来实现称量的。图2.1为电子分析天平的示意图。天平的秤盘通过支架连杆与线圈连接，线圈被置于磁场内。线圈中的电流产生磁场，支撑或悬浮支架连杆、秤盘及其负载和指示臂。调整电流，使秤盘空载时，指示臂处于零位置。当在秤盘上加载样品时，秤盘和指示臂向下移动，增加了击中零检测器光电池的光量，所增加的电流被放大并流入线圈，产生一个向上的电磁力，使秤盘返回其初始零位置，这样的装置被称为伺服系统。将秤盘和物体保持在零位置所需的电流与物体的质量成正比，并且易于测量、可数字化显示。电子分析天平的校准方法是使用标准质量的砝码调整电流，使其质量显示在屏幕上。

电子分析天平在使用过程中会受到所处环境温度、气流、震动、电磁干扰等因素的影响，因此要在稳定、干扰小的环境中使用。

Section 1　Analytical Balance

An analytical balance is the most crucial instrument for precise weighing in the experiment and can be categorized into mechanical and electronic types. Mechanical analytical balances can be subdivided into ordinary analytical balances, air-damped balances, semi-automatic electro-optical balances, automatic electro-optical balances and single pan balances. These balances all work on the lever principle, only differ in structure and method of use. The chief advantage of such balances lies in their intuitive structure; however, their components are complex, operation requirements are demanding, and the process is time-consuming.

In recent years, the analytical balance based on the lever principle has been gradually phased out and replaced by the electronic analytical balance that functions based on the principle of electromagnetic force balance. The electronic analytical balance boasts characteristics such as fast speed, no weights, digital readout, stable performance, convenience and accuracy. In addition, it also possesses functions like automatic calibration and taring control, overload display, fault alarm and capability for computer control and data logging.

Generally, an electronic analytical balance has a precision of 0.1 mg, namely it can weigh out 0.1 mg mass or distinguish 0.1 mg difference. The precision of a semi-micro electronic analytical balance is 0.01 mg. The precision of a micro electronic analytical balance is 0.001 mg, and that of an ultra-micro electronic analytical balance is 0.000 1 mg. The maximum capacity of electronic analytical balance is generally 100–200 g.

Ⅰ. Principle of Electronic Analytical Balance

The electronic analytical balance is based on the balance between the electromagnetic force and the gravity of the measured object. Figure 2. 1 shows a schematic diagram of an electronic analytical balance. The pan rides above a hollow metal cylinder that is surrounded by a coil that fits over the inner pole of a cylindrical permanent magnet. An electric current in the coil produces a magnetic field that supports or levitates the cylinder, the pan and indicator arm, and whatever load is on the pan. The current is adjusted so that the level of the indicator arm is in the null position when the pan is empty. Placing an object on the pan causes the pan and indicator arm to move downward, which increases the amount of light striking the photocell of the null detector. The increased current from the photocell is amplified and fed into the coil, creating a large magnetic field, which returns the pan to its original null position. A device such as this, in

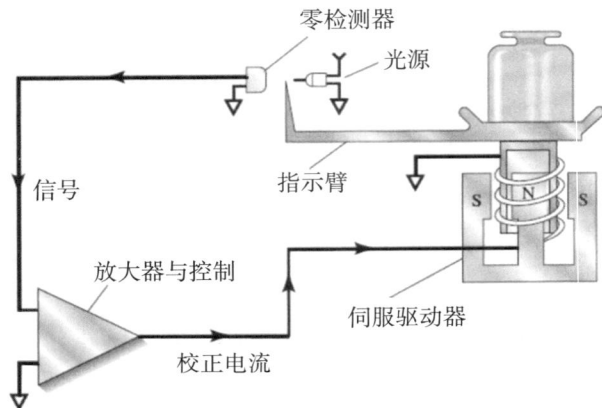

(a) 原理结构图   (b) 实物照片

图2.1   电子分析天平

二、电子分析天平的操作步骤

1. 调节水平

观察天平下部的水平泡，若水平泡偏离中心，则调整天平底部的水平调节脚，使水平仪内气泡位于圆环中央。每次移动了天平之后，均应重新调节天平。

2. 预热

为了确保测量结果准确，首次称量前，应按产品说明书的要求对天平进行预热（一般为30分钟），保证通电后的预热时间充分。

3. 开机

接通外电路，天平自检结束，然后单击开关键（"ON/OFF"键），出现字样"0.000 0 g"后，即可进行称量。

4. 校准

为了获得准确的称量结果，必须对电子天平进行校准，以适应当地的重力加速度，特别是首次使用天平称量之前及改变位置之后。称量工作中也应定期进行校准。以下是校准方法：

（1）准备好校准砝码，让秤盘空置，按住校准键（"CAL"键），至显示屏出现"CAL"字样再松开该键，所需要的校准砝码值会在显示屏上闪烁。

（2）把所需的校准砝码放在秤盘中央，天平会自动进行校准，当"0.000 0 g"闪烁时，移去砝码。

（3）在显示屏上短时间出现（闪现）"CAL done"字样，紧接着又出现"0.000 0 g"字样时，天平校准结束，回到称量工作状态，等待称量。

which a small electric current causes a mechanical system to maintain a null position, is called a servo system. The current required to keep the pan and object in the null position is directly proportional to the mass of the object and is readily measured, digitized, and displayed. The calibration of an electronic analytical balance involves using a standard mass and adjusting the current so that the mass of the standard is exhibited on the display.

The balance will be affected by ambient temperature, convection currents, vibration, electromagnetic interference and other factors when weighing, so it should be used in stable conditions.

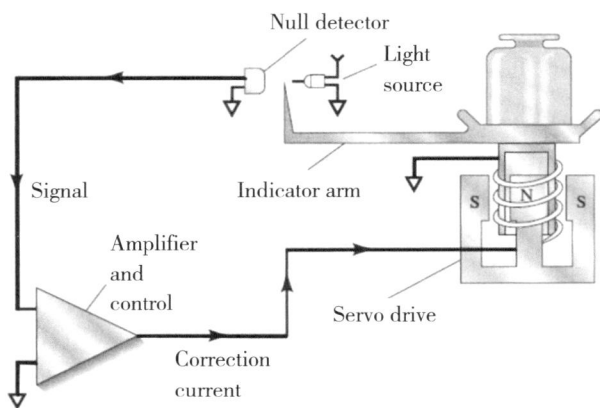

(a) Block diagram   (b) Photo of electronic balance

Figure 2. 1   Electronic analytical balance

II. Operating Procedures for Electronic Analytical Balance

1. Leveling

Check the leveling bubble at the bottom of the balance. If it is off-center, turn the 2 front feet of the balance until the air bubble is centered within the circle of the level indicator. Always level the balance again after any time it has been moved.

2. Preheating

To deliver exact results, the balance must warm up for a certain time (generally 30 min) according to the requirement of the product manual after initial connection to AC power or after a relatively long power outage.

3. Powering on

Connect the external circuit. Once the balance completes its self-check, click the power key ("ON/OFF" key). When the weighing mode shows "0.000 0 g", the weighing process may proceed.

4. Calibration

To obtain accurate weighing results, the electronic analytical balance must be calibrated to

5. 称量

根据称量对象和实验要求的不同，有以下三种称量方法：

（1）直接称量法。

此法用于称量某物体（如小烧杯、坩埚等）的质量。将天平电源开关打开，如有必要，按"Tare"键清零，然后将被称物直接放于天平秤盘正中央，读取天平显示屏数值即可。这种方法也适于称量洁净干燥、不易潮解或升华的固体试样。称量时，将洁净干燥的表面皿（或称量纸）放于天平秤盘正中央，按"Tare"键"去皮"或使读数归零，再将试样放于表面皿（或称量纸）上，读取天平显示屏数值即可。用该方法称量固体试样时，务必注意称量完成后，将所称取的试样完全转移到接收容器中，不得有损失。

（2）固定质量称量法。

此法也称增量法，用于称量固定质量的某试剂（如基准物质）或试样。这种方法称量的速度较慢，且适于称量不易吸潮、在空气中能稳定存在的试样，且试样应为粉末状或小颗粒状（最小颗粒应小于0.1 mg），以便调节其质量。固定质量称量法如图2.2所示，将一洁净的表面皿（或小烧杯）置于天平的秤盘上，按"Tare"键"去皮"，然后用食指轻轻敲击装有试样的药匙柄，慢慢地在表面皿上加试样至所需质量。称量时，若加入的试剂量超过了指定质量，则应重新称量。从试剂瓶中取出的试剂一般不应放回原试剂瓶中，以免玷污原试剂。操作时不能将试剂散落于表面皿（或小烧杯）以外的地方，称好的试剂必须定量地直接转入接收容器中。

图2.2   固定质量称量法

（3）递减称量法。

此法用于称量质量在一定范围内的试剂或试样，易吸水、易氧化或易与CO_2反应的试样可用此法称量。需平行多次称取某试剂时，也常用此方法。由于称取试样的质量是由两次称量之差求得，故也称差减法。

用此法称量时，先借助纸片从干燥器（或烘箱）中取出装有试样的称量瓶（注意：不要让手指接触称量瓶和瓶盖，称量瓶应处于室温），如图2.3（a）所示。将称量瓶置于秤盘正中央，关好天平门，称出称量瓶及试样准确质量（也可按"Tare"键，使其显

accommodate the local gravitational acceleration, particularly before the first use of the balance and after a change in location. Calibration should also be carried out periodically during the weighing process. Below shows the calibration process:

(1) Prepare calibration weights. Leave the pan empty. Press and hold the calibration key ("CAL" key) until the word "CAL" appears on the display. The required calibration weight value will flash on the display.

(2) Place the required calibration weight on the center of the pan. The balance is automatically calibrated, and when "0.000 0 g" flashes, remove the weight.

(3) When the display screen reads "CAL done" for a short time (flash), followed by "0.000 0 g," the calibration is finished, and the balance is back to the weighing state, ready for weighing.

5. Weighing

Different objects and experimental requirements adopt different weighing methods and operation steps. There are three commonly used weighing methods.

(1) Direct weighing.

This method is used to weigh the mass of an object, such as a beaker or a crucible. Turn on the balance. If necessary, press the "Tare" key, then place the weighed object directly on the center of the pan and take the reading. This method is also suitable for weighing solid samples that are clean and dry and not easily deliquescent or sublimated. When weighing, put a clean and dry watch glass (or weighing paper) on the center of the balance pan. Press the "Tare" key, and then put the sample on the watch glass (or weighing paper), read the weight. When using this method to weigh solid samples, it is essential to note that after weighing, the measured sample must be completely transferred to the receiving container without any loss.

(2) Fixed mass weighing.

Also known as weighing by addition, this method is utilized for weighing a certain fixed mass of a reagent (such as a primary standard) or sample. This method has a relatively slow weighing speed and is suitable for weighing samples that are not hygroscopic and can stably exist in the air. Moreover, the sample should be in powder or small granular form (the smallest particle should be lighter than 0.1 mg) to facilitate adjusting its mass. The fixed mass weighing method is depicted in Figure 2. 2. A clean watch glass (or small beaker) is placed on the pan of the balance, and the "Tare" key is pressed for tare removal. Subsequently, the handle of the spatula containing the sample is gently tapped with the index finger, and the sample is slowly added to the watch glass until the required mass is achieved. During the weighing process, if the amount of sample added exceeds the anticipated mass, reweighing is necessary. Generally, to avoid contamination, the reagent taken out of the reagent bottle must not be returned. During the operation, the reagent should not be scattered outside the watch glass (or small beaker), and the weighed reagent must be quantitatively and directly transferred into the receiver.

(3) Weighing by difference.

This method is employed for weighing reagents or samples within a certain mass range.

示"0.0000 g")。再将称量瓶取出，用另外一张小纸条捏住称量瓶盖子，打开称量瓶，在接收容器的上方，倾斜瓶身，用称量瓶盖轻敲瓶口上部使试样慢慢落入容器中，如图2.3（b）所示。当敲落的试样接近所需量时（一般称第2份时可根据第1份的体积估计），一边继续用瓶盖轻敲瓶口，一边逐渐将瓶身竖直，使黏附在瓶口上的试样落下，然后盖好瓶盖，把称量瓶放回天平秤盘，准确称出其质量。两次质量之差即为试样的质量（若先清了零，则显示值为试样质量）。若一次差减出的试样量未达到要求的质量范围，可重复操作，直至合乎要求。按此方法连续递减，可称取多份试样。

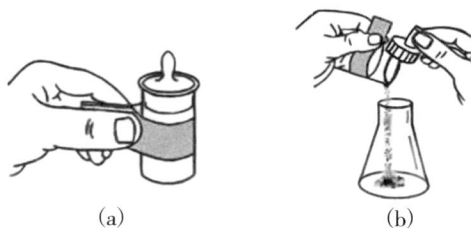

(a)　　　　(b)

图2.3　递减称量法

6. 天平复原

称量结束，按住开关键（"ON/OFF"键），至显示屏出现"OFF"字样再松开按键，关闭天平，切断电源，清洁、整理天平。

三、使用电子分析天平的注意事项

电子分析天平是一种精密的仪器，操作要细心，请从指导教师处获取并认真阅读所用型号天平的使用说明。不论品牌或型号，使用电子分析天平时都要遵守以下一般规则：

（1）开、关天平，放、取被称物，开、关天平门等，力度都要轻缓，切不可用力按压、冲击天平秤盘，以免损坏天平。

（2）清零和读取称量读数时，要关好天平门。要把称量读数立即记录在实验记录本中。

（3）称量物应放在天平秤盘的正中央。

（4）保护天平不受腐蚀。放置在秤盘上的容器应为非活性金属、非活性塑料、玻璃或玻璃状材料。

（5）对于热的或过冷的被称物，应置于干燥容器中直至其温度同天平室温度一致后才能进行称量。

Samples that are prone to water absorption, oxidation, or reaction with CO_2 can be weighed using this method. It is also used in weighing replicate samples. The sample mass is derived from the difference between two weighings, hence the name Weighing by Difference

When weighing, first take out the weighing bottle containing the sample from the desiccator (or oven) with a strip of paper (Note: Do not let fingers touch the weighing bottle and its cap; the weighing bottle should be at room temperature), as shown in Figure 2.3 (a). Place the weighing bottle on the center of the pan. Close the balance doors, and weigh the weighing bottle and its contents (or press the "Tare" key to read 0.000 0 g). Then take out the weighing bottle, hold the cap of the weighing bottle with a small piece of paper. Open the weighing bottle. Tilt the bottle over the receiving container. Tap the top of the weighing bottle with the cap to make the sample falling into the container slowly, as shown in Figure 2.3 (b). When the anticipated amount has been transferred (generally the amount of the second portion can be estimated according to that of the first portion), slightly turn up the bottle; meanwhile, keep tapping the bottle to make the sample particles adhering to the bottle mouth falling into the receiving container, and then cover the bottle. Put the weighing bottle back onto the pan, and read the mass. The difference between the two masses is the weight of the sample (if zeroing first, the displayed value is the weight of the sample). If the difference is out of the required mass range, repeat the operation until it reaches the required amount. Replicate samples can be weighed by successive weighing by difference.

Figure 2.2   Fixed mass weighing

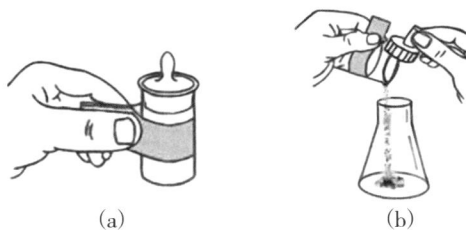

(a)   (b)

Figure 2.3   Weighing by difference

6. Recovering the balance

After weighing, press the "ON/OFF" key until it reads "OFF" on the display screen. Cut off the power supply, clean and tidy the balance.

III. Notes for Using an Electronic Analytical Balance

An electronic analytical balance is a delicate instrument that one must handle with care. Consult with the instructor for detailed instructions on weighing with the particular model of balance. Observe the following general rules for working with an analytical balance regardless of make or model:

(1) Be gentle and slow when turning on/off the balance, placing/taking the sample and opening/closing the balance door, etc. Do not press and hit the balance pan, so as not to damage the balance.

(2) Close the balance door when zeroing and reading the weighing. Record the readings

（6）在天平箱内放置变色硅胶干燥剂，当变色硅胶失效后应及时更换。

（7）注意保持天平、天平台和天平室的整洁和干燥。可在天平箱内放置小毛刷用于清除溢出的物质或灰尘。

（8）使用钳子、指垫或小纸条拿取干燥的容器，以防止水分转移到容器上。

（9）如果发现天平工作不正常，应及时向教师或实验工作人员报告，不要自行处理。

（10）称完后，应及时使天平还原，并在天平使用登记本上登记。

promptly in the laboratory notebook.

(3)　Center the object to be weighed on the pan as well as possible.

(4)　Protect the balance from corrosion. Objects to be placed on the pan should be limited to nonreactive metals, nonreactive plastics, and vitreous, or glasslike materials.

(5)　Always allow hot or super-cold objects to return to room temperature in a desiccator before weighing them.

(6)　Allochroic silica gel desiccant is placed in the balance case and should be replaced in time after failure.

(7)　Keep the balance and its case scrupulously clean. A hair brush is useful for removing spilled material or dust.

(8)　Use tongs, finger pads, or a glassine paper strip to handle dried objects to prevent transferring moisture to them.

(9)　Consult the instructor or laboratory staff if the balance appears to need adjustment. Do not handle it on one's own.

(10)　After weighing, restore the balance in time and register in the balance use log.

第二节　滴定分析仪器及其基本操作

　　在滴定分析中，用于准确测量溶液体积的玻璃仪器有滴定管、容量瓶、移液管和吸量管。正确使用这些玻璃仪器，是滴定分析最基本的操作技术。下面介绍滴定分析中常用的仪器及使用方法。

　　1. 滴定管

　　滴定管是滴定时用来准确测量流出的滴定剂体积的量器。常量分析用的滴定管容积为50 mL和25 mL，最小分度值为0.1 mL，读数可估计到0.01 mL。此外，还有容积为10 mL、5 mL、2 mL和1 mL的半微量和微量滴定管，最小分度值为0.05 mL、0.01 mL或0.005 mL。它们的形状各异。

　　滴定管由一个装滴定剂的刻度直管和一个控制滴定剂流动的阀门装置组成。根据阀门装置的不同，滴定管可分成两种类型：一种是酸式滴定管，下部带有磨口玻璃活塞，如图 2.4（a）所示；另一种是碱式滴定管，它的下端连接橡皮管，内放玻璃珠，橡皮管下端再连接一尖嘴玻璃管，见图 2.4（b）。只有当橡皮管变形时，滴定剂才能流过玻璃珠。

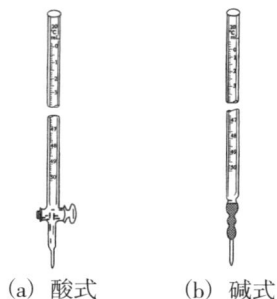

(a) 酸式　　(b) 碱式

图 2.4　滴定管

　　酸式滴定管只能用来盛放酸性、中性或氧化性溶液，不能盛放碱性溶液，因其磨口玻璃活塞会被碱性溶液腐蚀，放置久了会粘连住。碱式滴定管用来盛放碱性溶液，不能

Section 2   Apparatus and Operations for Precisely Measuring Volume

In titrimetric analysis, volume may be measured reliably with a buret, a volumetric flask or a pipet. The correct use of these tools serves as the foundation for titrimetric work. The tools and usage methods will be introduced in the following.

1. Buret

A buret is used for the accurate delivery of a variable amount of solution in titrations. The common capacities are 50 mL and 25 mL with a minimum graduation of 0.1 mL, and they can be read to the nearest 0.01 mL. In addition, there are semi-micro and micro burets with capacities of 10 mL, 5 mL, 2 mL and 1 mL and minimum graduations of 0.05 mL, 0.01 mL or 0.005 mL. They come in different shapes.

A buret consists of a calibrated tube to hold titrant plus a valve arrangement by which the flow of titrant is controlled. This valve is the principal source of difference among burets. An acid buret equipped with a glass stopcock for a valve relies on a lubricant between the ground-glass surfaces of stopcock and barrel for a liquid-tight seal [Figure 2.4 (a)]. A basic buret with simplest pinchcock valve consists of a close-fitting glass bead inside a short length of rubber tubing that connects the buret and its tip [Figure 2.4(b)]. Only when the tubing is deformed does liquid flow past the bead.

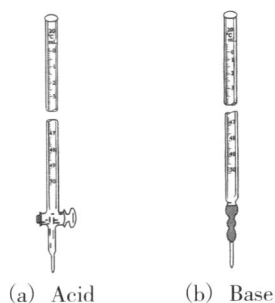

(a) Acid   (b) Base

Figure 2.4   Burets

Acid burets can only be used to hold acid, neutral or oxidizing solutions. Some solutions, notably bases, cause glass stopcocks to freeze when they are in contact with ground glass for long periods. Basic burets are used to release alkali liquor, but not oxidizing solutions such as $KMnO_4$, I_2 or $AgNO_3$, etc., to avoid corrosion of the rubber tube.

Most burets made in the last several of decades have Teflon valves. They look like acid

盛放氧化性溶液如$KMnO_4$、I_2或$AgNO_3$等，避免腐蚀橡皮管。

现在许多实验室常使用聚四氟乙烯酸碱两用滴定管，其形状与酸式滴定管相同，不同之处在于其旋塞是用聚四氟乙烯材料做成的，耐腐蚀、不用涂油、密封性好。

（1）滴定管的准备。

滴定管一般用自来水冲洗，零刻度以上部位可用毛刷刷洗，零刻度以下部位如不干净，则应采用洗液[①]清洗（碱式滴定管应除去橡皮管，用橡胶乳头将滴定管下口堵住）。污垢少时可加入约10 mL洗液，双手平托滴定管的两端，不断转动滴定管，使洗液润洗滴定管内壁，操作时管口对准洗液瓶口，以防洗液外流。洗完后，将洗液分别由两端放出。如果滴定管太脏，可以将洗液装满整根滴定管浸泡一段时间。为防止洗液流出，在滴定管下方可放一烧杯。最后用自来水、蒸馏水洗净。洗净后的滴定管内壁应完全被水均匀润湿而不挂水珠。如挂水珠，应重新洗涤。

滴定管洗涤好后，可在其中装入蒸馏水至零刻度以上，并垂直地夹在滴定管架上，静置几分钟，观察是否漏水。然后试着滴定一下，看能否灵活控制滴定速度。若滴定管漏水或操作不灵活，应进行下述处理：

对于酸式滴定管，应取下活塞，用滤纸擦干净活塞及塞座。用手指蘸取少量（切勿过多）凡士林，在活塞大头端涂极薄的一层（注意远离活塞孔），在塞座小端内涂少量，把活塞径直插入塞座内，向同一方向转动活塞（不要来回转），直到从外面观察到凡士林均匀透明为止。如果是滴定管的出口管尖被润滑油堵塞，可先用水充满全管，将出口管尖浸入热水中，温热片刻后，打开活塞，使管内的水流突然冲下，将溶化的油脂带出。最后用小孔胶圈套在活塞小头槽内，防止活塞滑出而损坏。

使用碱式滴定管前应检查橡皮管长度是否合适，是否老化变质。要求橡皮管内玻璃珠的大小合适，能灵活控制液滴。如发现不合要求，应重新装玻璃珠和橡皮管。

若为带聚四氟乙烯旋塞的通用型滴定管，则通过螺丝调节即可。

（2）装入溶液和排气。

首先将操作溶液摇匀，使凝结在瓶内壁上的液珠混入溶液。确保活塞已关闭。再用该溶液润洗已清洗的滴定管内壁三次，每次用10~15 mL溶液。用操作溶液洗滴定管时，要注意务必使操作溶液洗遍全管，并使溶液与管壁接触1~2 min，每次都要冲洗滴定管出口管尖，并尽量放尽残留溶液。重复这个步骤两次以上。应小心地将操作溶液直接倒

---

①铬酸洗液（$K_2Cr_2O_7$–浓H_2SO_4溶液）的配制：20 g的$K_2Cr_2O_7$，溶于40 mL水中，将360 mL浓H_2SO_4徐徐加入$K_2Cr_2O_7$溶液中（千万不能将水或溶液加入H_2SO_4中），边倒边用玻璃棒搅拌，并注意不要溅出，混合均匀，冷却后，装入洗液瓶备用。新配制的洗液为红褐色，氧化能力很强；洗液用久后变为黑绿色（可加入固体高锰酸钾使其再生），即说明洗液无氧化洗涤力。因浓H_2SO_4易吸水，应用磨口玻璃塞塞好。由于铬酸洗液是一种酸性很强的强氧化剂，腐蚀性很强，易烫伤皮肤，烧坏衣物，且铬有毒，所以使用时要注意安全和环境保护。

burets and are unaffected by most common reagents and require no lubricant.

(1) Preparing the buret.

Burets are generally washed with running tap water. The part above the zero mark can be washed with a brush; but if the part below the zero mark is dirty, it should be washed with a cleaning liquid, like chromic acid mixture [1] (The rubber tube of the basic buret should be replaced with a rubber nipple). Add about 10 mL of the cleaning liquid when there is slight dirt. Hold both ends of the buret flatly, and constantly roll the buret to allow the cleaning liquid to moisten the inner wall. During the operation, the buret should be held right above the washing bottle to prevent spillage. After washing, release the cleaning liquid from both ends of the buret. If the buret is extremely dirty, fill the buret with cleaning liquid and soak for a period of time. Place a beaker under the buret to collect any outflowing washings. Finally, it must be thoroughly rinsed with tap water and then with three or four portions of distilled water. Only clean glass surfaces support a uniform film of liquid. Inspect for water breaks and repeat the treatment if necessary.

Once the buret is clean, add distilled water to above the zero mark, and clamp it vertically to a buret stand. Let it sit for a few minutes and check for leakage. Then check if the stopcock can be freely controlled. If the stopcock leaks or is not flexible, the following treatment should be carried out:

For acid burets, remove the stopcock and clean it and its barrel with filter paper and dry both parts completely. Dip your finger in a small amount of vaseline, apply a very thin layer on the big end of the stopcock (using very little vaseline near the hole and taking care not to get any grease in the hole), lightly grease the small end of the barrel. Insert the stopcock into the barrel and rotate it vigorously in one direction with slight inward pressure until the area of contact between stopcock and barrel appears nearly transparent. If the tip of the buret is blocked with grease, you can first fill the whole tube with water, immerse the tip in hot water and warm for a moment, and open the stopcock to allow the water in the tube to rush down suddenly, carrying away the melted grease. Finally, a small rubber ring is set in the small head groove of the glass stopcock to prevent it from slipping out and damaging.

Before using the basic buret, check whether the length of the rubber tube is appropriate or whether there is aging deterioration, whether the size of the glass bead in the rubber tube is suitable and whether the droplets can be flexibly controlled. Reinstall the glass bead and rubber tube if necessary.

---

[1]Preparation of chromic acid cleaning solution ($K_2Cr_2O_7$-concentrated H_2SO_4 solution): Dissolve 20g of $K_2Cr_2O_7$ in 40 mL of water. Add 360 mL of concentrated H_2SO_4 slowly to the $K_2Cr_2O_7$ solution (do not add water or solution to H_2SO_4) with stirring, and pay attention to not splash, mix thoroughly, after cooling, put into the washing bottle for reserve. The newly prepared cleaning liquid is reddish-brown with strong oxidation ability. It becomes black green after it is used for a long time (add solid $KMnO_4$ to regenerate), indicating little oxidation and cleaning power. Concentrated H_2SO_4 is easy to absorb water, so the solution bottle should be plugged with grinding glass stopper. Because chromic acid mixture is a strong oxidizer with strong acidity, it is very corrosive, easy to irritate skin, burn clothes, and chromium is toxic, so pay attention to safety and environmental protection when using.

入滴定管中，不能用其他容器（如烧杯、漏斗等）转移溶液。倒入操作溶液时，关闭活塞，用左手大拇指、食指与中指持滴定管上端无刻度处，稍微倾斜，右手拿住细口瓶往滴定管中倒入操作溶液，让溶液沿滴定管内壁缓缓流下，倒入操作溶液至零刻度以上为止。为使溶液充满出口管（不能留有气泡），在使用酸式及通用型滴定管时，右手拿滴定管上部无刻度处，滴定管倾斜约30°，左手迅速打开活塞使溶液冲出，从而可使溶液充满全部出口管。如出口管中仍留有气泡，可重复操作几次。如仍不能使溶液充满，可能是出口管部分未洗涤干净，必须重新洗涤。对于碱式滴定管，应注意玻璃珠下方的洗涤。用操作溶液洗涤完后，将其装满溶液，垂直地夹在滴定管架上，左手拇指和食指放在稍高于玻璃珠所在的部位，并使橡皮管向上弯曲（见图2.5），出口管斜向上，往一旁轻轻提高挤捏橡皮管，使溶液从管口喷出，再一边捏橡皮管，一边将其放直，这样可排出出口管的气泡，并使溶液充满出口管。注意，放直橡皮管后再松开拇指和食指，否则出口管仍会有气泡。排尽气泡后，加入操作溶液使之在零刻度以上，再调节液面至零刻度或稍下处，记下初读数，读至0.01 mL。

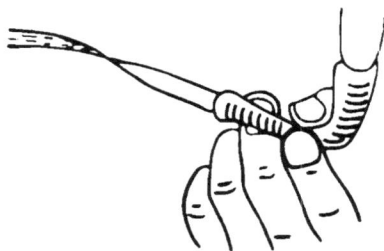

图2.5   碱式滴定管排气方法

（3）滴定管的使用。

①滴定管的操作。将滴定管垂直地夹于滴定管架上的滴定管夹上。

使用酸式和通用型滴定管时，用左手控制活塞，无名指和小指向手心弯曲，轻轻抵住出口管，大拇指在前，食指和中指在后，手指略微弯曲，轻轻向内扣住活塞，手心空握，如图2.6（a）所示。转动活塞时切勿向外（右）用力，以防顶出活塞，造成漏液。也不要过分往里拉，以免造成活塞转动困难，不能自如操作。

使用碱式滴定管时，左手拇指在前，食指、中指在后，三指尖固定住橡皮管中玻璃珠，挤橡皮管内玻璃珠的外侧（以左手手心为内），使橡皮管与玻璃珠之间形成一条缝隙，从而放出溶液，见图2.6（b）。注意不能捏玻璃珠下方的橡皮管，以免松开手时空气进入而形成气泡，也不要用力捏玻璃珠，或使玻璃珠上下移动。

②滴定操作。滴定通常在锥形瓶中进行，锥形瓶下垫白瓷板作为背景，右手拇指、

If the buret is equipped with a Teflon stopcock, the valve can be adjusted using the screw on the stopcock.

(2) Filling.

First swirl the titrant solution to catch up the droplets condensing on the inner wall of the bottle. Make certain the stopcock is closed. Add 10–15 mL of the titrant, and carefully roll and tilt the buret to wet the interior completely. Allow the liquid to drain through the tip. Repeat this procedure at least two more times. Note that the titrant should be carefully poured directly into the buret and should not be transferred through other containers (such as beakers, funnels, etc.). When pouring the titrant solution, hold the upper end of the buret with the thumb, index and middle fingers and tilt it slightly. Hold the solution bottle in the other hand. Pour the titrant into the buret, allowing the solution to flow down the inner wall of the buret gradually. Fill the buret well above the zero mark. For acid and universal burets, free the tip of air bubbles by tilting the buret and rapidly opening the stopcock and permitting small quantities of the titrant to pass. Repeat the operation if necessary. For the basic buret, after rinsing with the titrant solution, it is first filled with the titrant and clamped vertically on the buret stand. Place your left thumb and forefinger slightly above the glass bead and bend the tip upward (Figure 2. 5). Squeeze the glass bead gently to deform the rubber tube and make the solution out from the tip, and then put the rubber tube straight while kneading the glass bead, which can work air bubbles out and fill the tip with the titrant. Be sure that the rubber tube is straight before releasing the thumb and forefinger; otherwise, the tip will still have air bubbles. Finally, lower the level of the liquid just to or somewhat below the zero mark. Allow for drainage for about 1 min, and then record the initial volume reading, estimating to the nearest 0.01 mL.

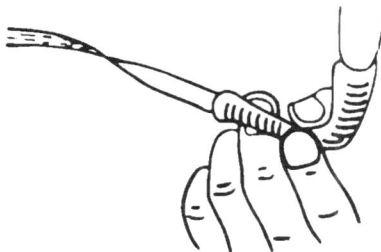

Figure 2. 5   Removing air bubbles inside the tip of a basic buret

(3) Using burets.

① Operation. Clamp the buret vertically on the buret stand.

When using acid and universal burets, manipulate the stopcock with the left hand by positioning the ring and small fingers slightly bent against the tip, the thumb at the front, and the index and middle fingers at the back, slightly bent to gently secure the stopcock inward, as illustrated in Figure 2. 6 (a). Do not push the stopcock outward (right) to avoid ejecting it and causing leakage. Do not go so far inward that the stopcock has trouble turning.

When using basic burets, position the left thumb at the front and the index and middle

食指和中指捏住瓶颈，瓶底离瓷板2~3 cm。调节滴定管高度，使其下端伸入瓶口约1 cm。左手按前述方法操作滴定管，右手用手腕的力量摇动锥形瓶，使瓶内液体按逆（或顺）时针方向做水平圆周运动，边滴加溶液边摇动锥形瓶，见图2.6（c）。

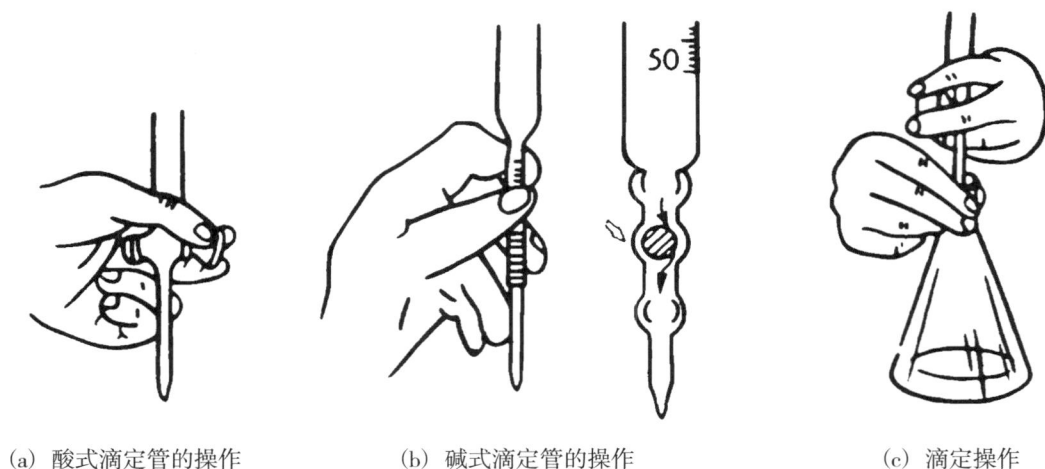

(a) 酸式滴定管的操作　　　　(b) 碱式滴定管的操作　　　　(c) 滴定操作

图2.6　滴定管的使用

在整个滴定过程中，左手一直不能离开活塞而任溶液自流。摇动锥形瓶时，应微动腕关节，使溶液向同一方向旋转，不要前后振动，以免溶液溅出，勿使瓶口碰到滴定管口，也不要使瓶底碰到白瓷板。一般在滴定开始时，溶液无可见的变化，滴定速度可稍快，一般为10 mL/min，呈"见滴成线"状，即每秒3~4滴。滴定到一定时候，滴落点周围出现暂时性的颜色变化。在离滴定终点较远时，颜色变化立即消失。临近终点时，变色甚至可以暂时地扩散到全部溶液，不过在摇动1~2次后变色完全消失。此时，应改为滴1滴，摇几下。等到必须摇2~3次，颜色变化才完全消失时，说明离滴定终点已经很近。微微转动活塞使溶液悬在出口管嘴上形成半滴，但未落下，用锥形瓶内壁将其沾下。然后将瓶倾斜把附于壁上的溶液洗入瓶中，再摇匀溶液。如此重复直至恰好出现达到终点时出现的颜色，而该颜色又不再消失为止。一般30 s内不再变色即达到滴定终点。

每次滴定最好都从读数0.00开始，也可以从0.00附近的某一读数开始，这样在重复测定时，使用同一段滴定管，可减小误差，提高精密度。

滴定完毕，弃去滴定管内剩余的溶液，不得倒回原瓶。用自来水、蒸馏水冲洗滴定管，并装入蒸馏水到刻度以上，用一小玻璃管套在管口上，保存备用。

③滴定管读数。滴定开始前和滴定结束后都要读取数值。滴定管读数前，应看看出口管嘴上是否挂着液珠。滴定后，若出口管嘴上挂有液珠，则无法准确确定滴定体积。读数时，从滴定管夹上取下滴定管，用右手大拇指和食指捏住滴定管上部无刻度处，使

fingers at the back to secure the glass bead within the rubber tube using the fingertips of these three fingers. Squeeze the outer side of the glass bead in the rubber tube （inner side facing the palm of the left hand） to create a gap between the tube and the glass bead and release the solution, as shown in Figure 2. 6 （b） . Be careful not to pinch the rubber tube under the glass bead, to avoid air bubbles when you let go the hand. Do not pinch the glass bead forcibly, or make the glass bead move up and down.

　② Titration. Titration is typically conducted in an Erlenmeyer flask, with a white porcelain plate placed beneath it as a backdrop, and the buret tip is positioned about 1 cm within the neck of the flask. Hold the neck of the Erlenmeyer flask with the right thumb, index and middle fingers, and the bottom of the flask is 2−3 cm from the porcelain plate. The left hand operates the buret as described above, and the right hand swirls the Erlenmeyer flask to move the solution in one direction. Keep swirling the Erlenmeyer flask while adding the titrant, as shown in Figure 2. 6(c).

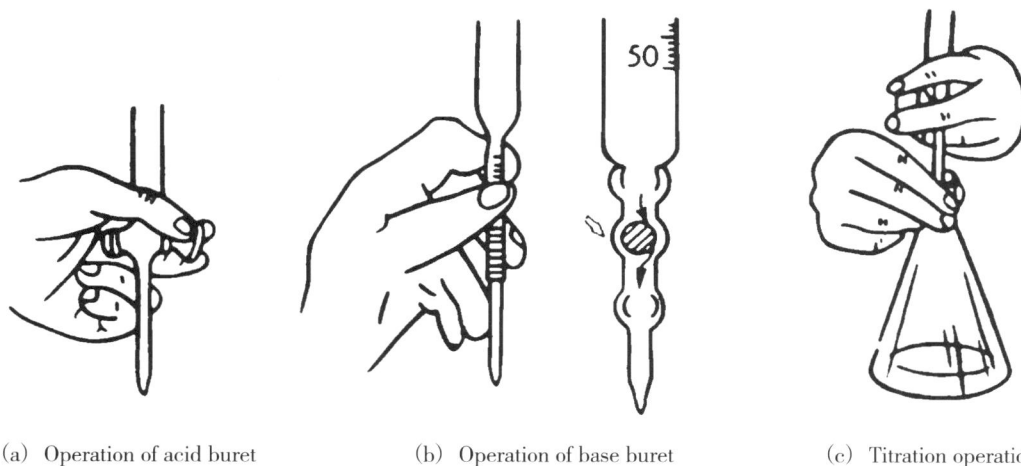

(a)　Operation of acid buret　　　(b)　Operation of base buret　　　(c)　Titration operation

Figure 2. 6　Operation of burets and titration

During the titration, the left hand must remain on the stopcock to regulate solution flow— never permit uncontrolled dispensing. Swirl the Erlenmeyer flask with a gentle wrist motion to rotate the solution unidirectionally. Prevent abrupt shaking (to avoid splashing), and maintain clearance between the flask mouth and buret tip, as well as between the flask base and white porcelain plate. Generally, at the beginning of titration, there is no discernible change in the solution, and the titration speed can be marginally faster, typically at 10 mL/min, in a state of "drops into line," that is, 3−4 drops /s. After a certain time, a temporary color change occurs in the vicinity where the titrant is added. Rapid disappearance of this color indicates it is far from the endpoint. In the vicinity of the endpoint, the color change can even temporarily spread throughout the solution, but after one or two swirls, it disappears completely. At this point, swirl the flask every one addition drop. When the color change is completely gone after swirling 2−3 times, it is close to the endpoint. Turn the stopcock slightly to let a half drop hanging （not

滴定管自然下垂，然后再读数。在滴定管中的溶液形成一个弯液面，无色或浅色溶液的弯液面下缘比较清晰，易于读数。读数时，使弯液面的最低点与分度线上边缘的水平面相切，视线与分度线上边缘在同一水平面上，以防产生误差。因为液面是球面，采用不同视角会得到不同的读数，见图2.7（a）。

为了便于读数，可在滴定管后衬读数卡。读数卡可用黑纸或涂有黑长方形（约3 cm×1.5 cm）的白纸制成。读数时，手持读数卡置于滴定管背后，使黑色部分在弯液面下约1 mm处，此时即可看到弯液面的反射层成为黑色，然后读此黑色弯液面下缘的最低点，见图2.7（b）。

在使用带有蓝色衬背的滴定管时，液面呈现三角交叉点，应读取交叉点与刻度相交之点的读数，见图2.7（c）。

颜色太深的溶液，如$KMnO_4$、I_2溶液等，很难看清楚弯液面，可读取液面两侧的最高点，此时视线应与该点持平。

必须注意，初始读数与终点读数应采用同一读数方法。刚刚添加完溶液或刚刚滴定完毕，不要立即调整零点或读数，而应等0.5～1 min，以使管壁附着的溶液流下来，使读数准确可靠。读数须准确至0.01 mL。在读取初始读数前，如果滴定管尖悬挂液滴，应该用锥形瓶外壁将液滴沾去。在读取终点读数前，如果出口管尖悬有溶液，此次读数不能用。注意在每次读数前，都要检查一下管壁内有没有挂水珠，管的尖嘴处有无悬液滴，管嘴内有无气泡。

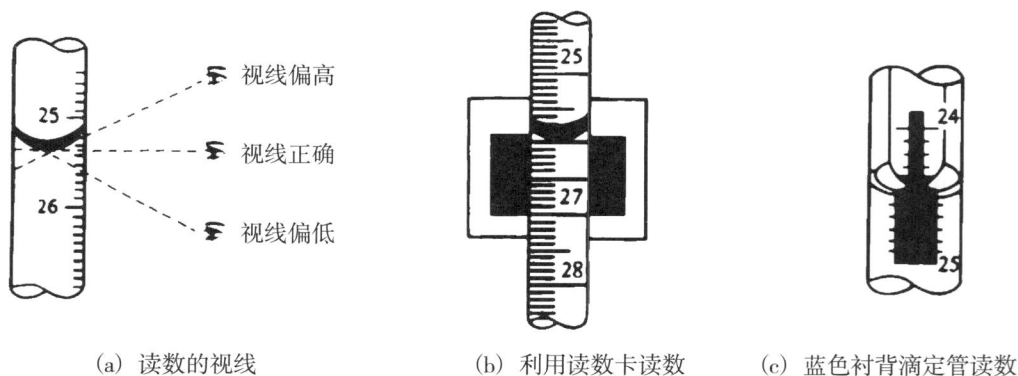

（a）读数的视线   （b）利用读数卡读数   （c）蓝色衬背滴定管读数

图2.7　滴定管读数

2. 容量瓶

容量瓶是细颈梨形的平底玻璃瓶，带有玻璃磨口塞或塑料塞。颈上有标线，表示在特定温度下（一般为20 ℃）当液体达到标线时瓶内液体的体积。容量瓶主要用于配制标

falling) on the buret tip. Dip it by touching the inside neck of the Erlenmeyer flask. Then tilt the flask to let the solution collect the half drop attached to the neck wall, and swirl the solution well. This is repeated until the color associated with the titration endpoint appears and persists. Generally, if the color change persists for 30s, the endpoint is considered reached.

It is advisable to begin each titration with a reading near 0.00, so that error can be reduced and precision is improved by using the same section of the buret in replicate measurements.

After a titration is complete, the unused titrant should never be returned to the original bottle but should be discarded. Rinse the buret with tap water and distilled water, and add distilled water to above the mark. Cover the top of the buret with a small glass tube, and store it for later use.

③ Reading. Take reading at the initial and end of the titration. Before reading, be sure that there is no liquid drop hanging on the tip of the buret. When reading, remove the buret from the buret stand, hold the top of the buret to allow it to naturally droop, and then read. The top surface of a liquid confined in a narrow tube exhibits a marked curvature, or called meniscus. It is common practice to use the bottom of the meniscus as the point of reference in calibrating and using volumetric equipment. In reading volumes, the eye must be at the level of the liquid surface to avoid an error due to parallax. Parallax is a condition that causes the volume to appear smaller than its actual value if the meniscus is viewed from above and larger if the meniscus is viewed from below as shown in Figure 2. 7(a).

A buret reading card with a black rectangle (approx. 3 cm × 1. 5 cm) can help you to take a more accurate reading. It is placed behind the buret, the upper limit of the black streak is about 1 mm below the meniscus, which darkens the meniscus and makes it easier to distinguish the bottom of the liquid surface, as shown in Figure 2. 7(b).

When a buret with blue backing is used, the liquid level shows a triangular crossing point and the reading should be taken at where the crossing point intersects the graduation as illustrated in Figure 2. 7(c).

For solutions with an overly deep color, such as $KMnO_4$ and I_2, it is quite difficult to clearly observe the bottom of the meniscus. In such cases, the highest points on both sides of the liquid surface can be read, and the line of sight should be level with the highest points.

Note to take the initial and final reading in a consistent manner. Just after the solution is added or the titration is completed, do not bring the liquid level to the zero point or take reading immediately. Instead, wait for 0.5-1 min to make certain the drainage film has caught up to the meniscus, so that the reading is accurate and reliable. Read the volume to the nearest 0.01 mL. Before taking the initial reading, if a drop is hanging on the buret tip, the droplet should be removed by touching the outer wall of the Erlenmeyer flask. Before taking the endpoint reading, if a droplet is hanging from the buret tip, this reading cannot be used. Prior to each reading, remember to check whether there are water droplets on the inner wall, whether there are drops hanging on the tip, and whether there are air bubbles inside the tip.

准溶液或试样溶液，也可以用于将一定量的浓溶液稀释成准确体积的稀溶液。通常有 25 mL、 50 mL、100 mL、250 mL、500 mL、1 000 mL等规格。

（1）容量瓶的准备。

在使用容量瓶前应先检查瓶塞是否漏水，再看标线位置离瓶口是否太近。容量瓶漏水则无法准确配制溶液，标线离瓶口太近则不便混匀溶液，因此都不宜使用。

检查漏水的方法是加自来水至标线附近，塞紧瓶塞。用食指按住塞子，将瓶倒立 2 min，如图 2.8（a）所示。如不漏水，将瓶直立，旋转瓶塞180°，塞紧，再倒立2 min，如仍不漏水，则可使用。

应将检验合格的容量瓶洗涤干净。洗涤方法、原则与洗涤滴定管相同。洗净的容量瓶内壁应均匀润湿，不挂水珠，否则必须重洗。

必须保持瓶塞与瓶子的配套，标以记号或用细绳、橡皮筋等把瓶塞系在瓶颈上，以防跌碎或与其他瓶塞混淆。

（2）容量瓶的操作。

用固体物质配制溶液时，准确称取一定量的固体物质，置于小烧杯中，加水或其他溶剂使其全部溶解（如果物质难溶，可盖上表面皿，加热溶解，但须放冷后才能转移），定量转移入容量瓶中。转移时，将玻璃棒伸入容量瓶中，使其下端靠在瓶颈内壁，上端不要碰瓶口，烧杯嘴要紧靠玻璃棒，使溶液沿玻璃棒和内壁流入，如图 2.8（b）所示。将溶液全部转移后，将玻璃棒稍向上提起，同时使烧杯直立，将玻璃棒放回烧杯。用洗瓶吸入蒸馏水吹洗玻璃棒和烧杯内壁，将洗涤液也转移至容量瓶中。如此重复洗涤至少3次。完成定量转移后，加水至容量瓶容积的一半左右，将容量瓶摇晃几周（勿倒转），使溶液初步混匀。继续加入溶剂并混匀，然后把容量瓶平放在桌上，慢慢加水到接近标线1 cm左右，等1～2 min，使黏附在瓶颈内壁的溶液流下。将细长滴管伸入瓶颈接近液面处，眼睛平视标线，加水至弯液面最低点与标线相切（注意：勿使滴管接触溶液，也可用洗瓶加蒸馏水至刻度）。立即塞上干燥的瓶塞，按图2.8（c）握持容量瓶的姿势（对于容积小于100 mL的容量瓶，只用左手操作即可），将容量瓶倒转，使气泡上升到顶。将容量瓶立正后，再次倒立振荡，如此重复10次左右，使溶液混合均匀。最后放正容量瓶，打开瓶塞，使其周围的溶液流下。重新塞好瓶塞，再倒立振荡1～2次，使溶液全部充分混匀。

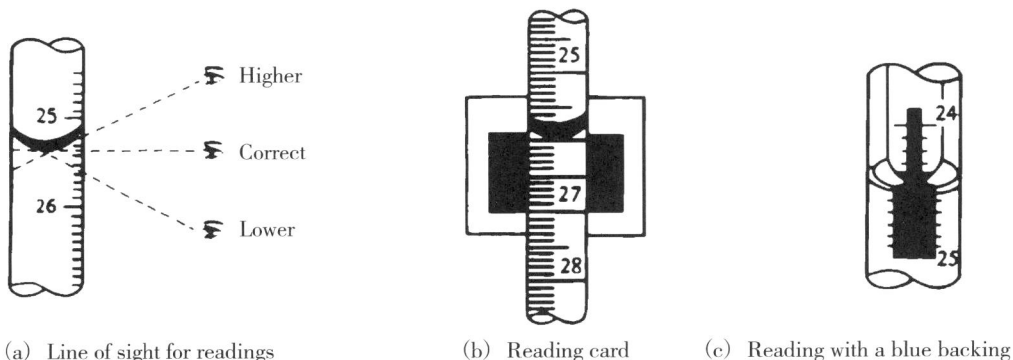

(a) Line of sight for readings   (b) Reading card   (c) Reading with a blue backing

Figure 2. 7   Reading a buret

2. Volumetric flask

A volumetric flask is a flat-bottomed glass bottle with ground glass or plastic stoppers. It is characterized by a bulb and a long neck. The flask is designed to accommodate the desired volume of the liquid up to the mark (a line etched on the neck) at a specific temperature (typically 20 ℃). It is used for preparing standard or sample solutions and for diluting a concentrated solution to a fixed volume. The flasks are available in various sizes like 25 mL, 50 mL, 100 mL, 250 mL, 500 mL and 1 000 mL, etc.

(1) Preparation.

Volumetric flasks should be first checked for leakage and whether the marking line is too close to the bottle opening. Leaky flasks cannot provide accurate solution volumes. If the mark is too close to the bottle mouth, it is inconvenient to thoroughly mix the solution. Therefore, neither is suitable for use.

Check for leaks by adding tap water to the mark and capping the flask. Press the stopper with your index finger and stand the bottle upside down for 2 min, as shown in Figure 2. 8 (a). In the event of no leakage, upright the bottle, rotate the stopper by 180° , fasten it tightly, and invert the bottle for another 2 min. If no leakage is detected, then the flask can be employed.

Before use, volumetric flasks should be thoroughly cleaned and rinsed with distilled water. The inner wall of the washed volumetric flask should be uniformly moistened without water drops. Otherwise it must be rewashed.

The stopper must be kept paired with its corresponding volumetric flask, marked or secured to the neck of the flask using a string or rubber band to prevent misplacement or potential confusion with other stoppers.

(2) Operation.

When preparing a solution from a solid substance, weigh an exact amount of the solid into a small beaker. Add water or other solvents to dissolve it all (If the substance is insoluble, cover a watch glass, heat to dissolve the solute, and allow the solution to cool to room temperature). Transfer this solution quantitatively into the volumetric flask, as described in the following. When transferring, insert a stirring rod into and against the neck of the volumetric flask, and

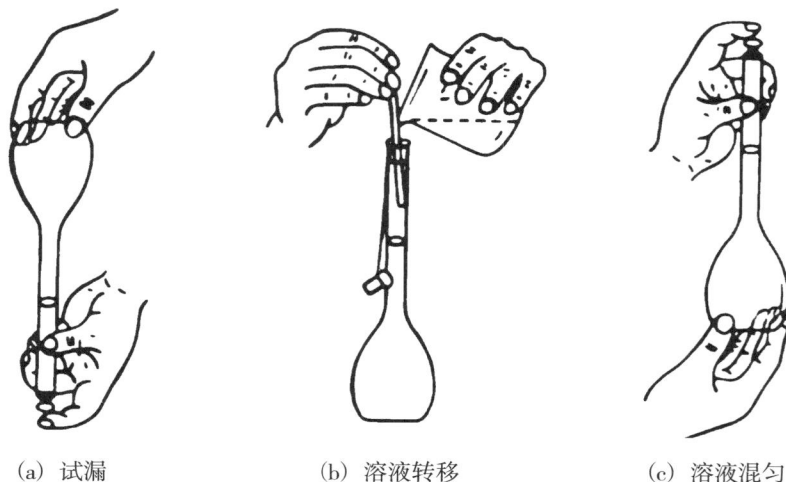

(a) 试漏   (b) 溶液转移   (c) 溶液混匀

图 2.8 容量瓶的操作示意图

稀释溶液时，用移液管移取一定体积的溶液于容量瓶中，加蒸馏水至标度刻线，然后按上述方法混匀溶液。

注意不能用手掌握住瓶身，以免体温造成液体膨胀，影响容积测量的准确性。热溶液应冷却至室温后，才能注入容量瓶中，否则可造成瓶塞粘住，无法打开。配好的溶液如需长期保存，应转移到试剂瓶中，不要将容量瓶当作试剂瓶使用。使用完毕应立即用水洗干净，若长期不用，在洗净擦干磨口后，用纸片将磨口隔开。此外，容量瓶不能加热，更不能在烘箱中烘烤。如需使用干燥的容量瓶，可在洗净后用乙醇等有机溶剂清洗，然后晾干或用电吹风的冷风吹干。

3. 移液管和吸量管

移液管用来准确移取一定体积的溶液。它是中间有一较大空腔的细长玻璃管，管颈上部刻有一标线，如图 2.9 （a） 所示，膨大部分标有它的容积和标定时的温度。在标明的温度下，若使溶液的弯液面与移液管标线相切，再让溶液按一定的方法自由流出，则流出溶液的体积与管上标明的体积相同。常见的移液管有 5 mL、10 mL、25 mL、50 mL 等规格。

吸量管是带有分刻度的玻璃管，如图 2.9 （b） 所示。它用于准确移取所需不同体积的液体。常用的吸量管有 1 mL、2 mL、5 mL、10 mL 等规格。吸量管量取溶液的准确度不如移液管。

use this stirring rod to direct the flow of liquid from the beaker into the flask, as shown in Figure 2. 8 (b). After all the solution has been transferred, with the stirring rod, tip off the last drop of liquid on the spout of the beaker. Rinse both the stirring rod and the interior of the beaker with distilled water and transfer the washings to the volumetric flask as well. Repeat the rinsing process at least three times. After the solute is transferred, fill the flask about half full and swirl the contents to hasten solution. Add more solvent and again mix well. Bring the liquid level almost to the mark, and allow time for drainage (1–2 min). Then with your eye at the same level with the graduation mark, use a slender dropper to make any necessary final additions of solvent until the lowest point of the meniscus aligns precisely with the line (Note: Avoid letting the dropper touch the solution. Alternatively, use the wash bottle to add distilled water up to the mark). Immediately seal the flask with a dry stopper and hold it as shown in Figure 2.8(c) (for flasks under 100 mL, only the left hand is needed). Invert the flask to allow bubbles to rise to the top. Return the flask to an upright position, then invert and shake it again, repeating this process about 10 times to ensure thorough mixing. Finally, place the flask upright, open the stopper to let any surrounding solution drain down, reseal it, and invert-shake once or twice more to achieve complete homogeneity.

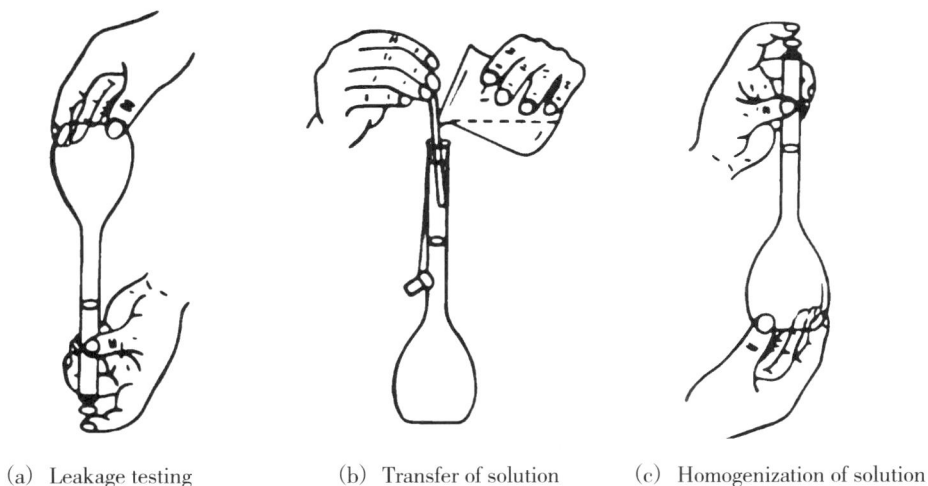

(a) Leakage testing   (b) Transfer of solution   (c) Homogenization of solution

Figure 2. 8 Operation of a volumetric flask

When diluting a solution, pipet a certain volume of the solution into a volumetric flask, add distilled water to the marking line, and then mix the solution.

It is essential to avoid holding the bottle body with your palm to prevent body heat from inducing thermal expansion of the liquid, which may influence the accuracy of the volume. Hot solutions should be cooled to room temperature before being poured into the volumetric flask; otherwise, the stopper will be stuck and cannot be opened. If the prepared solution needs to be stored for a long time, it should be transferred to a storage bottle that either is dry or has been thoroughly rinsed with several small portions of the solution from the flask. Do not use the volumetric flask as the reagent bottle. Flush it with water immediately after use. If it is not in use for a long time, separate the stopper and the grinding mouth with a piece of paper after washing

(a) 移液管   (b) 吸量管

图2.9   移液管和吸量管

移液管和吸量管的使用方法如下：

（1）洗涤。

有油污的移液管一般先用洗耳球吸取铬酸洗液洗涤，让洗液布满全管，静置1～2 min，将洗液放回原瓶。用洗液洗涤后，沥尽洗液，用自来水充分冲洗，再用蒸馏水润洗3次，润洗的水应从管尖放出。洗好的移液管必须达到内壁与外壁的下部完全不挂水珠的状态。清洗完毕，将其放在干净的移液管架上。

（2）移取溶液。

移取溶液前，先用吸水纸将管尖内、外的水除去，否则水滴引入会改变溶液的浓度。然后用待取溶液润洗3次，以确保所移取的操作溶液浓度不变。润洗的操作方法是：用洗耳球吸取待取溶液至刚入移液管的膨大部分（注意勿使溶液回流，以免稀释及玷污待取溶液），立即用右手食指按住管口，将管横过来，用两手的拇指和食指分别拿住移液管的两端，转动移液管并使溶液布满全管内壁，当溶液流至距上口2~3 cm时，将管直立，从管尖放出溶液。

移取待取溶液时，将移液管尖插入液面下1～2 cm。管尖不应伸入液面太深，以免管外壁黏附过多的溶液；也不应伸入太浅，否则液面下降后会吸空。当管内液面借洗耳球的吸力而慢慢上升时，移液管尖应随着管内液面的下降而下降。当管内液面升高到刻度以上时，移去洗耳球，迅速用右手食指堵住管口（食指最好是潮而不湿），将管上提，离开液面。将移液管原伸入溶液的部分，贴容器内壁转两圈，尽量除去管尖外壁黏附的溶液。然后将容器倾斜成45°左右，竖直移液管，管尖紧贴容器内壁，稍松右手食指（使食指造成的压力减小，但注意食指不要离开管口），用右手拇指及中指轻轻捻转管身，使液面缓慢而平稳地下降，直到溶液弯液面的最低点与刻度上边缘相切，视线与刻度上边缘在同一水平面上，立即停止捻动并用食指按紧管口，保持容器内壁与移液管尖接触，以除去吸附于移液管尖的液滴。取出移液管，立即插入接收容器中，使接收容器倾斜约45°而管直立，松开食指，让管内溶液自由地顺壁流下（如图2.10所示），待液

and drying. In addition, volumetric flasks should not be heated, or dried in an oven. To get a dry volumetric flask, you can wash it with ethanol and other organic solvents after cleaning, then air dry or blow dry with the cool air of a hair dryer.

3. Pipet

Pipets permit the transfer of accurately known volumes from one container to another. There are two common types of pipets, the volumetric, or transfer, pipet and the measuring or graduated pipet. Variations of the latter are also called clinical, or serological, pipets.

The volumetric pipet [see Figure 2.9(a)] is a thin glass tube with a bulb shape in the middle. It is used for accurate measurements since it is designed to deliver only one volume and is calibrated at that volume. At the indicated temperature, if the level of the meniscus exactly lines up with the mark on the neck, and the solution is allowed to drain freely in a certain way, the volume of the delivered liquid is the same as the volume marked on the pipet. Common volume of the pipets may be 5 mL, 10 mL, 25 mL, 50 mL, etc.

The measuring pipet is a straight-bore pipet that is marked at different volume intervals, as shown in Figure 2.9 (b). It is used for accurate delivery of liquids of different volumes required. The commonly used pipets have 1 mL, 2 mL, 5 mL, 10 mL and other total capacities. These are not as accurate because nonuniformity of the internal diameter of the device will have a relatively larger effect on total volume than is the case for volumetric pipets.

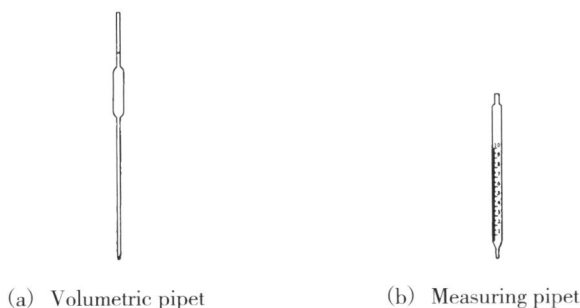

(a) Volumetric pipet   (b) Measuring pipet

Figrue 2. 9   Pipets

The following part introduces the operating procedures for volumetric pipets and measuring pipets.

(1) Cleaning.

Draw detergent solution or chromic acid mixture to a level 2 to 3 cm above the calibration mark of the pipet. Drain this solution and then rinse the pipet with several portions of tap water. Inspect for film breaks, and repeat this portion of the cleaning cycle if necessary. Finally, fill the pipet with distilled water to perhaps one third of its capacity and carefully rotate it so that the entire interior surface is wetted. Repeat this rinsing step at least twice.

(2) Measuring an aliquot.

Before pipetting, sop up the water inside and outside the tip with a absorbent paper. Then

面下降到管尖后，停靠15 s左右，同时左右旋转移液管，最后再移出移液管。这时，管尖部位仍留有少量溶液，对此，除特别注明"吹"字的以外，此管尖部位留存的溶液是不能被吹入接收容器中的。

图2.10　放出溶液操作

使用吸量管移取溶液的操作与用移液管移取溶液的操作基本相同。实验中，要尽量使用同一支吸量管，以免带来误差。

移液管和吸量管用完后应立即用自来水冲洗，再用蒸馏水冲洗干净，放在管架上备用。

rinse 3 times with the solution to be delivered to ensure constant concentration. The rinsing operation is as follows: draw the solution just into the bulb of the pipet with a pipet bulb (Do not allow the solution to flow back to dilute and contaminate the solution). Immediately press the flat end of the pipet with the index finger, turn the pipet horizontally, and hold both ends of the pipet. Rotate the pipet to coat the entire interior surface with the solution. After the solution flows to 2–3 cm from the top, hold the pipet upright to drain solution through the tip.

When pipetting the solution, insert the tip of the pipet 1–2 cm beneath the liquid surface. The tip should not be inserted too deeply into the liquid, as excessive solution may adhere to the outer wall of the pipet; nor should it be inserted too shallowly, otherwise the pipet may suck in air when the liquid level drops. After the liquid is drawn and went up in the pipet slowly, the pipet tip should descend along with the dropping liquid level in the container. When the liquid level in the tube rises above the volume mark, quickly remove the bulb and place your index finger firmly on the end of the pipet (index finger is best damp but not wet). Lift the pipet out of the solution, and rotate twice the inserted part of the pipet against the inner wall of the container to remove as much of the solution adhering to the outer wall of the tip as possible. Then tilt the solution container at an angle of approximately 45°. Hold the pipet upright with its tip against the inner wall of the container neck. Slightly loosen the index finger (reducing the pressure on the pipet opening but be careful not to remove the finger from the opening). Gently twist the pipet with your thumb and middle fingers to lower the liquid level slowly and steadily until the bottom of the meniscus hits the volume mark when viewed at eye–level. Immediately stop twisting and press your index finger firmly on the end of the pipet so no liquid leaks out. Touch the tip once to the side of the container to collect droplets. Pull out the pipet and immediately place the tip of the pipet against the wall of the receiving container tilted at an angle of approximately 45° (as shown in Figure 2. 10). Slowly allow the liquid to drain from the pipet. Keep the flow slow so that no droplets cling to the inside of the pipet. When the solution stops flowing, touch the pipet to the side of the receiving container for about 15 s, roll the pipet left and right at the same time to remove any hanging drops. Finally remove the pipet. Do not blow out the remaining solution. The pipet has been calibrated to deliver the appropriate amount of solution with some remaining in the tip.

The operation of a measuring pipet is basically the same as that of a transfer pipet. In the experiment, try to use the same pipet to minimize errors.

The pipet should be washed with tap water immediately after use, then rinsed with distilled water, and placed on the pipet rack for later use.

Figure 2. 10　Delivering an aliquot

第三节   酸度计

1. 酸度计简介

酸度计（以下简称pH计）是一种采用氢离子选择性电极测量液体pH值的广泛使用的电化学分析仪器，其基本原理是：将一个连有内参比电极的可逆氢离子指示电极和一个外参比电极同时浸入某一待测溶液中而形成原电池，在一定温度下产生一个内外参比电极之间的电池电动势。这个电动势与溶液中氢离子活度有关，而与其他离子的存在基本无关。仪器测量该电动势的大小就转化为待测溶液的pH值，并显示出来。

实验中为了操作方便，常常把连有内参比电极的氢离子指示电极和外参比电极复合在一起构成复合pH电极。复合pH电极的基本结构如图2.11所示。复合pH电极在第一次使用或在长期停用后再次使用前应在$3 \text{ mol} \cdot \text{L}^{-1}$ KCl溶液中浸泡24 h以上，使其活化。平时可浸泡在$3 \text{ mol} \cdot \text{L}^{-1}$ KCl溶液中保存。

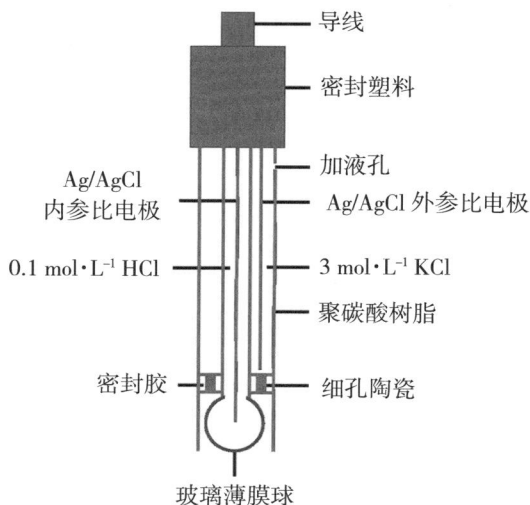

图 2.11   复合pH电极

Section 3   pH Meter

1. Introduction

An acidity meter （hereinafter referred to as a pH meter） is a widely used electrochemical analytical instrument for measuring the pH value of a liquid using a hydrogen ion selective electrode. Its working principle is as follows： A reversible hydrogen ion indicator electrode connected to an internal reference electrode and an external reference electrode are simultaneously immersed in a certain solution to be tested to form a galvanic cell， generating an electromotive force between the internal and external reference electrodes at a certain temperature. This electromotive force is related to the activity of hydrogen ions in the solution and is essentially unrelated to the presence of other ions. The instrument measures the magnitude of this electromotive force and finally converts it into the pH value of the solution to be tested and displays it.

In experiments， for operational convenience， the hydrogen ion indicator electrode connected to the internal reference electrode and the external reference electrode are often combined to form a single-probe， or combination， variety. The basic structure of the combination pH probe is depicted in Figure 2. 11. The combination pH probe should be immersed in $3 \ mol \cdot L^{-1}$ KCl solution for more than 24 h for activation before being used for the first time or again after long-term disuse. Usually， it can be stored by soaking in a $3 \ mol \cdot L^{-1}$ KCl solution.

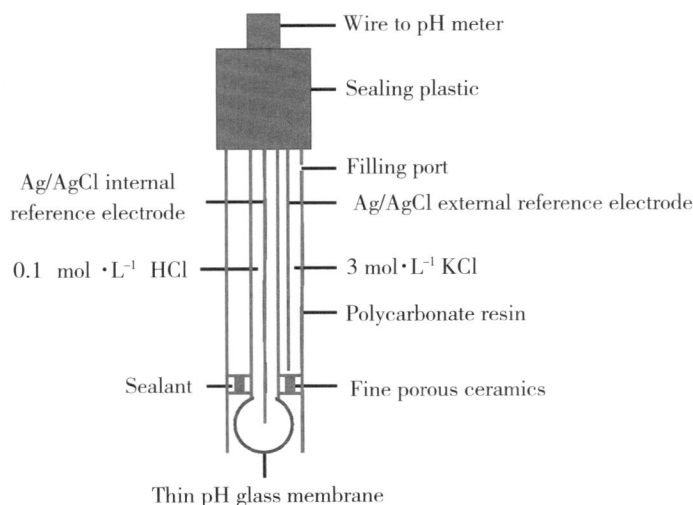

Figure 2. 11   Combination pH probe

2. PHS-3C型pH计

pH计型号较多，它们的结构、功能和使用方法大同小异。实验室中广泛使用的PHS-3C型pH计是一种精密数字显示pH计，其测量范围宽，重复误差小。PHS-3C型pH计由主机（包括液晶显示屏和按键）、复合电极及多功能支架组成（如图2.12）。

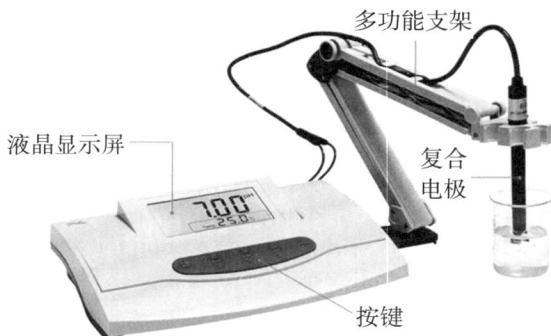

图2.12　PHS-3C型pH计（上海雷磁）

主机上有五个按钮，分别是pH/mV（选择）、定位、斜率、温度和确认按钮。

测量溶液pH的操作步骤如下：

（1）检查pH计的接线是否完好。接通电源，按下背面的电源开关，预热30 min后方可使用。

（2）取下复合电极上的电极套，注意不要将电极套中的KCl溶液撒出或倒掉。用蒸馏水冲洗复合电极头部，用吸水纸吸干残留水分。

（3）定位。在测量之前，首先对pH计进行校准，一般采用两点定位校准法，具体步骤如下：

①打开电源开关，按"pH/mV"按钮，使仪器进入pH测量状态。

②用温度计测量被测溶液的温度，读数，如25 ℃。按"温度"键设定测量值25 ℃，然后按"确认"键，回到pH测量状态。

③打开电极套管，用蒸馏水冲洗复合电极头部，用吸水纸仔细将复合电极头部吸干，将复合电极放入pH为6.86的标准缓冲溶液，使溶液淹没电极头部的玻璃球，轻轻摇匀，待读数稳定后，按"定位"键，使显示值为该溶液25 ℃时的标准pH值6.86，然后按"确认"键，回到pH测量状态。

④将复合电极取出，洗净、吸干，放入pH为4.00（或9.18）的标准缓冲溶液中，摇匀，待读数稳定后，按"斜率"键，使显示值为该溶液25 ℃时标准pH值，按"确认"键，回到pH测量状态。

2. PHS–3C pH Meter

There are numerous models of pH meters, and their structures, functions, and usage methods are largely similar. The PHS–3C type pH meter, which is widely employed in laboratories, is a precise digital-display pH meter featuring a broad measurement range and small repetitive errors. The PHS–3C type pH meter is composed of a mainframe and a probe (combination electrode), as shown in Figure 2. 12.

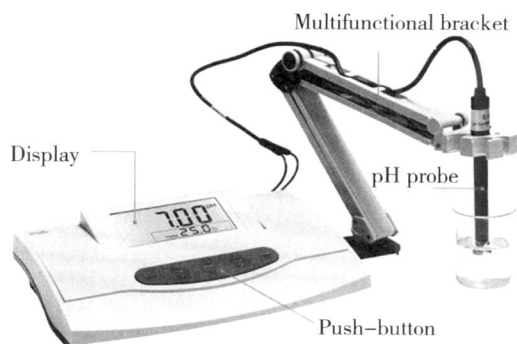

Figure 2. 12   PHS–3C pH meter (LEI–CI, Shanghai)

There are five buttons on the mainunit, namely: pH/mV (selection), positioning, slope, temperature and confirmation buttons.

The procedure for measuring the pH of a solution is as follows:

(1) Check if the pH meter is properly connected. Press the power switch on the back and allow it to preheat for 30 min before use.

(2) Remove the electrode cap from the combination probe, and be careful not to spill or dump the KCl solution inside the cap. Rinse the electrode glass tip with distilled water and blot the residual moisture with filter paper.

(3) Positioning. Before measurement, the pH meter should be calibrated first. Generally, the two-point positioning calibration method is adopted. The specific steps are as follows:

①Switch on the power supply and press the "pH/mV" button to enable the instrument to enter the pH measurement state.

②Measure the temperature of the tested solution using a thermometer and take the reading, e. g. , 25 ℃. Press the "Temperature" button to set the measured value as 25 ℃, and then press the "Confirmation" button to return to the pH measurement state.

③Remove the electrode cap, rinse the electrode tip with distilled water, and carefully blot the electrode head with a lint-free tissue. Insert the combination probe in a pH 6. 86 standard buffer. Make sure the glass bulb tip of the electrode is completely dipped into the solution, and stir (swirl) the solution gently. When the reading is stabilized, adjust the "Positioning" button to read the pH at 6. 86, then press the "Confirmation" button to return to the pH

⑤仪器的校正到此完成，可进行pH的测量。需要注意的是：校正好仪器后，不应再按"定位"及"斜率"键。若不小心触动了这些键，则不要按"确认"，而是按"pH/mV"键使仪器重新进入pH测量状态，这样就不需要再进行校正。一般情况下，每天校正一次即可。

（4）校正过程结束后，进入测量状态。用蒸馏水清洗电极，将复合电极放入盛有待测溶液的烧杯中，轻轻摇动，待读数稳定后，记录读数。若被测溶液与用于校正的溶液的温度不同，则先按"温度"键使仪器显示被测溶液的温度，再按"确认"键，进行pH测量。

完成测试后，移走溶液，用蒸馏水冲洗复合电极，吸干，套上套管，关闭电源，并进行仪器使用情况登记。

measurement state.

④Take out the electrode, clean and blot it dry, then place it in the standard buffer solution with a pH of 4. 00 (or 9. 18) . Swirl the solution well. After the reading stabilizes, press the "Slope" button to display the standard pH value at 25 ℃. Press the "Confirmation" key to return to the pH measurement state.

⑤ The calibration of the meter is hereby completed and it is now ready to take pH measurements. It should be noted that the "Positioning" and "Slope" of the instrument should not be pressed again after calibration. If these keys are accidentally touched, do not press "Confirmation" key. Instead, press the "pH/mV" key to make the instrument re-enter the pH measurement , thus eliminating the need for re-calibration. Generally, calibration is required at least once a day.

(4) After the calibration, enter the measurement mode. Clean the electrode with distilled water. Place the combination probe in a beaker containing the test solution, swirl it gently, and record the reading after it stabilizes. If the temperature of the test solution is different from that at which the calibration is performed, press "Temperature" to adjust it to that of the test solution, and then press "Confirmation" to obtain the pH reading.

After the test is accomplished, remove the solution, rinse the electrode with distilled water, blot it dry, put on the cap, switch off the power supply, and register the use of the instrument.

第四节   分光光度计

分光光度计是利用物质对不同波长的光的选择性吸收现象，对物质进行定性和定量分析的仪器。分光光度计可分为红外、紫外-可见、可见光等几类，有时也称为分光光度仪或光谱仪。可见光分光光度计用于可见光吸光光度法测定，较普遍使用的有721B型、722型和7220型，它们主要由图2.13中所示的五部分组成。

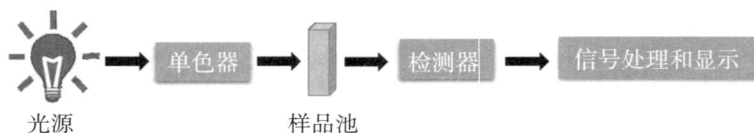

图2.13   分光光度计的主要部件

722S型分光光度计的外形及操作键如图2.14所示。

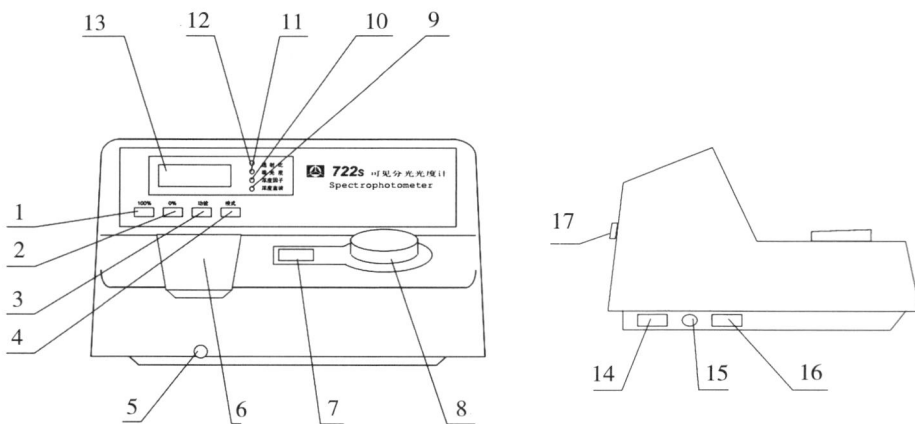

1. ↑ /100%键；2. ↓ /0%键；3. 功能键；4. 模式键；5. 试样槽架拉杆；6. 样品室；7. 波长指示窗；8. 波长调节旋钮；9. "浓度直读"指示灯；10. "浓度因子"指示灯；11. "吸光度"指示灯；12. "透射比"指示灯；13. 显示屏；14. 电源插座；15. 熔丝座；16. 总开关；17. RS232C 串行接口插座。

图2.14   722S型分光光度计外形图

Section 4   Spectrophotometer

A spectrometer or spectrophotometer is a device that measures the selective absorption of light by matters to obtain their qualitative and quantitative information. It can be classified into several types such as infrared, ultraviolet-visible and visible spectrophotometers. A visible spectrophotometer is employed in the determination of matters absorbing visible light. The commonly used models include 721B, 722 and 7220, which essentially consist of the following basic components as shown in Figure 2. 13.

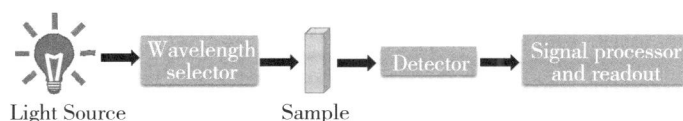

Figure 2. 13   Main components of a spectrophotometer

The appearance and operation keys of a 722S spectrophotometer are shown in Figure 2. 14.

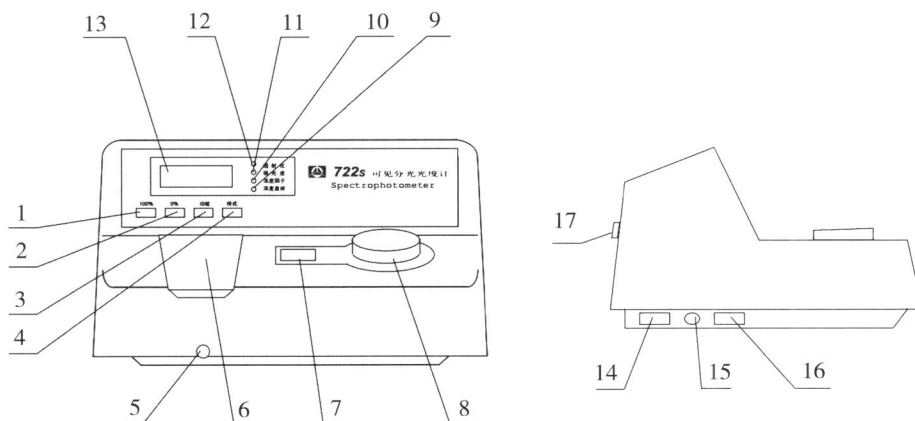

1. ↑ /100% Key; 2. ↓ /0% Key; 3. Function Key; 4. Mode Key; 5. Pull Rod of the Sample Slot Frame; 6. Sample Chamber; 7. Wavelength Indicator Window; 8. Wavelength Adjustment Knob; 9. "Concentration Direct Reading" Indicator Lamp; 10. "Concentration Factor" Indicator Lamp; 11. "Absorbance" Indicator Lamp; 12. "Transmittance" Indicator Lamp; 13. Display Window; 14. Power Socket; 15. Fuse Holder; 16. Main Switch; 17. RS232C Serial Interface Socket.

Figure 2. 14   722S spectrophotometer

其使用方法如下：

（1）仪器预热。接通电源，打开电源开关，打开样品室门，让仪器预热至少30 min，使仪器进入热稳定工作状态。

（2）测定透射比。按"模式"键，使"透射比"指示灯亮。调节波长旋钮至所需值，将装有参比溶液和待测溶液的比色皿置于样品池架中（注意：比色皿透明的面朝向入射光，手拿毛玻璃面），关上样品室门。将参比溶液拉至光路中，按"100%"键，使显示屏显示为"100.0"。打开样品室门，看显示屏是否显示"0.00"，若不是则按"0%"键，使显示屏显示为"0.00"。重复此两项操作，直至仪器显示稳定。然后将待测溶液依次拉入光路，读取各溶液的透射比。注意每当改变波长时，都应重新用参比溶液校正透射比"0.00"和"100%"。

（3）测定吸光度。在用参比溶液调好透射比T"100%"和"0.00"后 [见第（2）步]，按"模式"键直至"吸光度"指示灯亮，再将待测溶液依次拉入光路，在显示屏上读出各溶液的吸光度。通过测定标准溶液和未知溶液的吸光度A，绘制$A-c$工作曲线，根据未知溶液的吸光度可从工作曲线上找到对应的浓度值。作图时应合理选取横坐标与纵坐标数据单位比例，使图形接近正方形，工作曲线位于对角线附近。

（4）浓度直读。当对象分析规程比较稳定时，在标准曲线基本过原点的情况下，可不必采用步骤较复杂的标准曲线法而直接采用浓度直读法定量，本方法仅需配制一种浓度在用户要求定量浓度范围 2/3 左右的标准样品。操作如下：在第（2）步用参比溶液调好透射比T"100%"和"0.00"后，按"模式"键，使"浓度直读"指示灯亮，将标准溶液拉入光路，按↑或↓键使读数达已知含量值（或含量值的 $10n$ 倍），置入未知样品溶液，读出显示值即含量值（或含量值的 $10n$ 倍）。

（5）还原仪器。仪器使用完毕，关闭电源，拔下电源插头，取出比色皿，洗净，使仪器复原。然后盖上防尘罩，并进行仪器使用情况登记。

Its usage method is as follows:

(1) Preheating: Connect the power supply, turn on the power switch, and open the sample chamber. Allow the instrument to warm up for at least 30 min to enter a thermally stable operating state.

(2) Determination of transmittance: Press the "Mode" key to illuminate the "Transmittance" indicator. Adjust the wavelength knob to the desired value, place the cuvettes containing the reference and the test solutions in the sample chamber (Note: the transparent side of the cuvette should face the incident light, and hold the ground glass side by hand), and close the sample chamber door. Pull the reference solution into the beam path and press the "100%" key to make it display "100. 0". Open the sample chamber and check whether it reads "0.00". If not, press the "0%" key to make it display "0.00". Repeat these two operations until the reading is stable. Then, successively pull the sample solutions into the optical path and read the transmittance of each solution. Note that whenever the wavelength is changed, the transmittance "0.00" and "100%" should be recalibrated using the reference solution.

(3) Determination of absorbance: After adjusting the transmittance (T) to "100%" and "0.00" with the reference solution (see step 2), press the "Mode" key until the "Absorbance" indicator is illuminated. Subsequently, pull the solution to be tested successively into the optical path, and read the absorbance of each solution on the display screen. By measuring the absorbance (A) of the standard and unknown solutions, a standard curve of $A - c$ is constructed, and the corresponding concentration value can be found from the standard curve according to the absorbance of the unknown. When graphing, the data unit ratio of the abscissa and ordinate should be selected reasonably to make the graph close to a square, and the standard curve should be located near the diagonal.

(4) Direct reading of concentration: When the analytical procedure is stable and the standard curve basically passes through the origin, the direct reading of concentration for quantification can be adopted instead of the more complicated standard curve method. This method only requires the preparation of a standard sample with a concentration approximately 2/3 of the quantitative concentration range requested by the user. The operation is as follows: After adjusting T to "100%" and "0.00" using the reference solution as in step 2, press the "Mode" key to illuminate the "Concentration Direct Reading" indicator. Pull the standard solution into the optical path. Press the $\uparrow$ or $\downarrow$ key to bring the reading to the known concentration value (or $10n$ times of the known value). Place the unknown sample solution in the light path, and read its concentration (or $10n$ times of its concentration value).

(5) Restoring the instrument: After using the instrument, switch off the power, unplug the power plug, take out the cuvettes, clean them, and restore the instrument to its original state. Then cover it with a dust cover and make a registration of the instrument's usage.

第三章 定量分析基本操作实验

Chapter 3　Basic Operating Experiments in Quantitative Analysis

实验一　分析天平的使用

【实验目的】

（1）学会使用分析天平快速、准确称量样品质量。

（2）掌握两种称量方法：固定质量称量法和递减称量法（减量法）。

（3）培养准确、整齐、简明地记录原始实验数据的习惯，不可涂改数据，不可将测量数据记录在实验记录本以外的任何地方。

【实验原理】

准确称量物体的质量可用分析天平，目前常用电子分析天平，电子分析天平是基于电磁力平衡的原理工作的。秤盘与通电线圈相连接，置于磁场中，将被称物品置于秤盘上，因重力向下，线圈上就会产生一个电磁力，与重力大小相等、方向相反。传感器输出电信号，由此产生的电信号通过模拟系统后，将被称物品的质量显示出来。常用的电子分析天平最大载荷为100~200 g，分度值（感量）可达 ± 0.01~ ± 1 mg。

【仪器与试剂】

仪器：电子分析天平、称量瓶（内装沙子）、50 mL 烧杯、纸条、小药匙。

【实验步骤】

（1）调平：调整地脚螺栓高度，使水平仪内空气气泡位于圆环中央。

（2）预热：天平在初次接通电源或长时间断电之后，至少需要预热30 min。

（3）开机：按开关键直至全屏自检，显示天平型号，当天平显示归零时，就可以进行称量了。

（4）去皮：将一个50 mL的烧杯（或有折痕的称量纸）放在天平秤盘上。关上天平门，面板显示烧杯（或称量纸）的质量。按"去皮"按钮，显示器将再次显示为0.000 0 g。

（5）称量：

①固定质量称量法。

称取0.500 0 g沙子，称量方法有以下两种：

Experiment 1　Use of Analytical Balance

【Objectives】

(1) To learn to use the analytical balance to weigh the sample mass rapidly and accurately.

(2) To master two weighing methods: fixed mass weighing and weighing by difference.

(3) To systematically and accurately record the original experimental data in a clear and organized manner.

【Principle】

An object's mass is precisely measured using an analytical balance. The most common type of balance is an electronic balance in which the balance pan is placed over an electromagnet. The sample to be weighed is placed on the sample pan, displacing the pan downward by a force equal to the product of the sample's mass and the acceleration due to gravity. The balance detects this downward movement and generates a counterbalancing force using an electromagnet. The current needed to produce this force is proportional to the object's mass. A typical electronic analytical balance has a capacity of 100–200 g and can measure mass to the nearest ± 0.01 to ± 1 mg.

【Apparatus and Chemicals】

Apparatus: Analytical balance, weighing bottle (filled with sand), 50 mL beaker, paper strip, small spatula.

【Procedure】

(1) Leveling: Do check the level indicator bubble before weighing. The two balance feet serve as leveling screws.

(2) Preheating: The balance should be preheated for at least 30 min after the initial power supply is switched on or after a long power outage.

(3) Switching on: Turn the balance on by pressing "ON/OFF" button. The display lights up for several seconds, then resets to 0.000 0.

(4) Taring: Put a 50 mL beaker (or creased, small weighing paper) on the pan. Close the sliding glass doors. It displays the mass of the beaker. Press "TARE" button to cancel out the weight of the container or paper. The display will again read 0.000 0 g.

(5) Weighing:

①Fixed mass weighing.

a.用小药匙从称量瓶将试样慢慢加到小烧杯中，直到天平显示0.500 0 g，关上天平门，待读数稳定后，看读数是否仍为0.500 0 g。若所称重量小于该值，可继续加试样；若显示的量超过该值，则需重新称量。

b.取一张纸条，从纸条上撕下一小片纸，一手用纸条绕拿称量瓶，另一只手隔着小纸片捏拿称量瓶盖（注意：不能直接用手碰称量瓶）。把称量瓶放在烧杯上方，倾斜称量瓶，用瓶盖轻拍称量瓶敲出沙子。当转移的样品量达到预期值时，慢慢将称量瓶竖直，与此同时，不断轻拍称量瓶，使粘在瓶口上的样品落入烧杯。

至少称量3份0.500 0 g的沙子。

②递减称量法（减量法）。

a.递减称量法（下称减量法）即先称称量瓶（带盖）和样品的总质量，然后小心地从称量瓶中倒入少量样品到接收容器（如烧杯）中，再盖上称量瓶盖子，重新称量样品和称量瓶的总质量。两者之差即为转移到接收容器中的样品质量。注意：要用纸条或钳子拿称量瓶，不能直接用手接触称量瓶和盖子，因为指纹会影响称量质量。

b.用减量法称量0.3~0.4 g沙子。把称量瓶放在天平秤盘上，关上天平门，按"清零"键，使其显示0.000 0 g。取出称量瓶，将部分试样轻敲至小烧杯中，再称量，看天平读数是否在–0.4 ~–0.3 g范围内。若敲出量不够，则继续敲出，直至读数在此范围内，并记录数据。若敲出量超过0.4 g，则需重新称量。重复上述操作，称取第2份及第3份试样。

每次递减时，可根据称量瓶中试样的量或前一次所称试样的体积来判断敲出多少试样较合适，这样有助于提高称量速度。

【数据记录与处理】

将数据记录在表3.1、表3.2中。

表3.1　固定质量称量法

项目	序号		
	Ⅰ	Ⅱ	Ⅲ
m/g			

表3.2　减量法

项目	序号		
	Ⅰ	Ⅱ	Ⅲ
称量瓶倒出试样，m/g			

Weigh out 0.500 0 g of sand, and there are two weighing methods as follows:

a.Use a small spatula to transfer the sample from the weighing bottle to the small beaker gradually until the balance indicates 0.500 0 g. Close the balance door and wait for the reading to stabilize. Check if the reading remains 0.500 0 g. If the measured amount is less than this value, continue adding the sample; if the displayed amount exceeds this value, reweighing is necessary.

b.Take a strip of paper and tear off a small piece from it. With one hand, hold the weighing bottle wrapped with the paper strip, and with the other hand, pinch the weighing bottle cap through the small piece of paper (Note: Do not touch the weighing bottle directly with your hand). Place the weighing bottle above the beaker, tilt the weighing bottle, and gently tap the weighing bottle with the cap to make the sand come out of the bottle. When the anticipated amount of sample has been transferred, slowly upright the weighing bottle and, meanwhile, keep tapping the bottle gently to make the sample adhering to the bottle mouth fall into the beaker.

Weigh at least three 0.500 0 g portions of sand.

②Weighing by difference.

a.Weighing by difference involves weighing the weighing bottle, sample, and cap on the analytical balance, then dispensing a small amount of sample by carefully pouring some sample from the weighing bottle into a second container. Put the cap back on and reweigh the weighing bottle. The difference between these two masses is the amount of sample transferred to the flask or beaker. You should handle the sample (weighing) bottle and cap only with paper strip (or tongs) to avoid fingerprints, which can affect the masses.

b.Weigh 0.3–0.4 g of sand by difference. Position the weighing bottle on the pan of the balance, close the balance door, and press the "Zero" key to display 0.000 0 g. Take out the weighing bottle, gently tap a portion of the sample into a small beaker, and then reweigh. Check if the balance reading is within the range of −0.4 to −0.3 g. If the tapped-out amount is insufficient, continue tapping until the reading falls within this range, and record the data. If the tapped-out amount exceeds 0.4 g, a fresh weighing is necessary. Repeat the aforementioned operation to weigh the second and third samples.

During each difference, it is possible to determine the appropriate amount of sample to be tapped out based on the quantity of the sample in the weighing bottle or the volume of the previously tapped-out sample. This can facilitate an increase in the weighing speed.

【Data Recording and Processing】

Record the experimental data in Table 3.1 and Table 3.2.

Table 3.1   Fixed Mass Weighing

Item	No.		
	I	II	III
m/g			

Table 3.2   Weighing by Difference

Item	No.		
	I	II	III
Mass of sample dispensed from the weighing bottle, m/g			

【注意事项】

（1）不要用手直接拿要称量的物质，应该用干净的纸、钳子或镊子来拿。

（2）不要将化学品直接放在天平秤盘上，而要用容器（称量瓶、烧杯）或称量纸盛放。如果有药品溅出，要马上用软毛刷将溢出的化学物质刷干净。

（3）待称量物质必须放在天平秤盘的中央。

（4）所有要称量的物体或样品在称量之前必须与房间保持热平衡，天平内如果存在温度梯度就会产生对流，微弱气流作用于天平秤盘会引起称量误差。

（5）样品质量不要超过天平的最大载荷。

（6）称量和读数时，必须关上天平门，因为气流会使天平不稳定。

（7）记录天平读数到0.000 1 g。

（8）在烘箱中干燥后的样品应保存在干燥器中，以防止其重新吸收空气中的水分。

（9）取出的多余试剂应弃去，不要放回原试剂瓶中，不然会污染原瓶中的试剂。

【思考题】

（1）用减量法称量时，如何操作称量瓶？

（2）是否可以用手直接拿称量瓶进行称量，为什么？

（3）固定质量称量法和递减称量法各有何优缺点？分别在什么情况下选用这两种方法？

【Notes】

(1) Never handle objects to be weighed with the fingers. A piece of clean paper or tongs or tweezers should be used.

(2) Chemicals are never placed directly on the weighing pan. A weighing boat, beaker or paper is used under the chemical. If any chemicals spill, use a soft brush to clean up the spilled chemicals immediately.

(3) Center the load on the pan as much as possible.

(4) All objects or samples to be weighed must be in thermal equilibrium with the room before a weighing can be attempted. Temperature gradients within the balance chamber can result in convection currents, and the resultant draft against the pan will lead to erroneous weighings.

(5) Do not exceed the maximum weight limit of the balance.

(6) Always close the balance chamber door before making the weighing. Air currents will cause the balance to be unsteady.

(7) Always record weights to the nearest 0.000 1 g.

(8) Samples dried in an oven should be stored in a desiccator to prevent them from reabsorbing moisture from the atmosphere.

(9) Chemicals should never be returned to the stock bottle, as this may contaminate the chemicals in the bottle.

【Questions】

(1) How to manipulate a weighing bottle when weighing by difference?

(2) Is it acceptable to handle the weighing bottle directly with your hands? Why?

(3) Comment on the advantages and disadvantages of each of the two weighing methods used. Under what circumstances will each method be preferred?

实验二   容量瓶、移液管和滴定管的使用

【实验目的】

掌握容量瓶、移液管和滴定管的操作技术。

【实验原理】

在滴定分析中，规范地使用容量器皿及准确测量溶液的体积是保证获得良好分析结果的重要因素。为此，必须学习正确地使用容量瓶、移液管和滴定管等容量仪器的方法。

【仪器与试剂】

仪器：容量瓶（100 mL、250 mL）、移液管（25 mL）、滴定管、锥形瓶、烧杯、量筒、玻璃棒、小滴管、洗耳球。

试剂：蒸馏水、$KMnO_4$（固体）。

【实验步骤】

1. 容量瓶的使用和定量转移

（1）称取0.4 g固体$KMnO_4$，置于一洁净的50 mL烧杯中。注意：过量化学试剂应舍弃，不能送回原试剂瓶，以免污染瓶内试剂。

（2）加入约20 mL蒸馏水，缓慢搅拌溶解$KMnO_4$，制成溶液。此溶液为近饱和的溶液，小心操作，耐心等待固体完全溶解。

（3）将溶液定量转移到100 mL容量瓶中：将溶液沿玻璃棒注入容量瓶，溶液转移后，应将烧杯沿玻璃棒微微上提，同时使烧杯直立，避免沾在杯口的液滴流到杯外，再把玻璃棒放回烧杯。接着，用洗瓶洗涤烧杯内壁和玻璃棒，将洗液全部转移入容量瓶。

（4）重复洗涤烧杯和玻璃棒并转移洗液至容量瓶，直到烧杯中红色褪去。注意并记录将$KMnO_4$溶液从烧杯定量转移到容量瓶（烧杯中红色褪去）所需的洗涤次数。步骤（3）至（4）的操作过程称为定量转移。

（5）定量转移后，加水稀释，当加水至约大半瓶时，摇动容量瓶（不能倒置）使溶

Experiment 2   Use of Volumetric Flasks, Pipets and Burets

【Objectives】

To master the operation techniques of volumetric flasks, pipets and burets.

【Principle】

Correct use of volumetric glassware and accurate measurement of solution volume are essential for optimum accuracy in a volumetric determination, so it is necessary to learn how to use volumetric flasks, pipets and burets properly.

【Apparatus and Chemicals】

Apparatus: volumetric flask (100 mL, 250 mL), pipet (25 mL), buret, Erlenmeyer flask, beaker, measuring cylinder, glass rod, dropper, rubber suction bulb.

Chemicals: distilled water, $KMnO_4$(solid).

【Procedure】

1. Correct Use of the Volumetric Flask and Making Quantitative Transfers

(1) Weigh 0.4 g solid $KMnO_4$ in a 50 mL beaker on an electronic balance. Note that chemicals should never be returned to a stock bottle, as this may contaminate the bottle.

(2) Dissolve the solid $KMnO_4$ in the beaker using about 20 mL of distilled water. Stir gently to avoid loss. This is nearly a saturated solution, and some care is required to dissolve the crystals completely.

(3) Quantitatively transfer the solution to a 100 mL volumetric flask. Pour it down the glass rod, and then touch the rod to the spout of the beaker to remove the last drop. Add more water to the beaker, stir, and repeat the procedure.

(4) Repeat the procedure until no trace of the color of the $KMnO_4$ remains in the beaker. Note and record the number of washings that is required to quantitatively transfer the $KMnO_4$ solution from the beaker to the flask. It is called "quantitative transfer" which performed in step (3) – (4).

(5) After quantitative transfer, add water to dilute the solution. When the flask is filled to approximately a half of its capacity, swirl the volumetric flask (do not invert it) to initially mix the solution. Continue adding water until the liquid level is about 0.5 cm below the calibration mark. Use a small dropper to add distilled water dropwise until the liquid surface is tangent to the calibration line. Replace the stopper, press it firmly with the index finger,

液初步混匀，继续加蒸馏水至离刻度线约0.5 cm处，用小滴管逐滴加入蒸馏水至液面与标线相切，盖好容量瓶瓶塞，用食指压住塞子，其余四指握住瓶颈，另一手（五只手指）将容量瓶托住并反复倒置，摇荡，使溶液完全均匀，此操作称为定容。

（6）反复倒置容量瓶约10次，直到溶液完全均匀，保存溶液以备移取溶液实验使用。

2. 移取溶液

使用滴定管或移液管量取一定体积的溶液时，管内溶液必须与所移取的溶液主体具有相同的组分和浓度，下面的操作旨在说明如何清洗移液管和移取溶液。

（1）用洗耳球吸取$KMnO_4$溶液，使之充满移液管，随后放出溶液。

（2）在一洁净的50 mL烧杯中，加入蒸馏水，插入移液管，用洗耳球吸入蒸馏水至移液管约1/3处，倾斜移液管至近水平并旋转几周润洗整个内壁，弃去润洗液。不要将移液管完全装满水，这样的操作既浪费水，又耗时低效。

（3）重复吸水进入移液管并润洗内壁直至红色褪去，注意并记录红色褪去所需要的最少润洗次数。润洗方法正确的话，三次应该足够了。

（4）再次吸入$KMnO_4$溶液，然后像之前一样用蒸馏水润洗。这一次，将润洗液收集在一个量筒中，确定红色褪去所需的润洗水的最小体积。在润洗过程中，50 mL烧杯中的水是否被$KMnO_4$溶液污染了？如果出现淡红色，则表明烧杯中的水被污染，那么就要更小心地重复这个步骤。

（5）移取25 mL $KMnO_4$溶液于250 mL容量瓶中。

（6）小心缓慢地加蒸馏水，将溶液稀释至刻度，尽量不搅动瓶内液体。

（7）反复倒置和摇动容量瓶来混合溶液，注意并记录使$KMnO_4$溶液颜色均匀分布所需要的倒置和摇动次数。

（8）用步骤（7）容量瓶中的溶液润洗移液管3次，然后移取25 mL溶液到锥形瓶中。

3. 滴定管的使用和读数

（1）向滴定管中加入蒸馏水至近零刻度处，并置于滴定管架上。

（2）静置30 s后，读取初始体积。不要将滴定管中的液面精确地调整到0.00 mL，这样做只会带来偏差并浪费时间。

（3）放出5 mL水于250 mL锥形瓶中，等待至少30 s，然后读取最终体积。前后两次读数之差即为锥形瓶中的溶液的体积。在实验笔记本上记录滴定管的最终读数，然后请指导教师也读取出最终读数，两个读数之差应在0.01 mL内。请注意，滴定管读数中的最后一个数字是滴定管上两个连续0.1 mL标记之间距离的估计。

（4）重新装满滴定管，读取新的初始读数。从滴定管放出30滴水于锥形瓶中，读取

and grasp the neck of the flask with the other four fingers. With the remaining hand (using all five fingers) , hold the volumetric flask, invert it repeatedly, and shake vigorously to ensure the solution is fully homogeneous. This operation is referred to as " volume adjustment" or "making up to volume."

(6) Repeat until the solution is completely homogeneous; about 10 inversions and shakings are required. Save the solution for the next part.

2. Delivering an Aliquot

Whenever a buret or pipet is used to deliver a measured volume of solution, the liquid it contains before measurement should have the same composition and concentration as the solution to be dispensed. The following operations are designed to illustrate how to rinse and fill a pipet and how to deliver an aliquot of solution.

(1) Use a suction bulb, fill a pipet with the solution of $KMnO_4$ and let it drain.

(2) Draw a few milliliters of distilled water from a 50 mL beaker into the pipet, rinse all internal surfaces of the pipet, and discard the rinse solution. Do not fill the pipet completely; this is wasteful, time-consuming, and inefficient. Just draw in a small amount, tilt the pipet horizontally, and turn it to rinse the sides.

(3) Repeat the process of drawing water in and rinsing the pipet until the red color is removed. Determine the minimum number of such rinsings required to completely remove the color of $KMnO_4$ from the pipet. If your technique is efficient, three rinsings should be enough.

(4) Again fill the pipet with $KMnO_4$ solution, and proceed as before. This time, determine the minimum volume of rinse water required to remove the color by collecting the rinsings in a graduated cylinder. In the rinsing operations, was the water in the 50 mL beaker contaminated with $KMnO_4$ solution? If a pink color shows that it was, repeat the exercise with more care.

(5) Pipet 25 mL of the $KMnO_4$ solution into a 250 mL volumetric flask.

(6) Carefully dilute the solution to the volumetric flask, trying to mix the contents as little as possible.

(7) Mix the solution by repeatedly inverting and shaking the flask. Note the effort that is required to disperse the color of $KMnO_4$ uniformly throughout the solution.

(8) Rinse the pipet with the solution in the volumetric flask. Pipet a 25 mL aliquot of the solution into an Erlenmeyer flask.

3. Using a Buret and Reading a Buret

(1) Mount a buret in a buret stand, and fill the buret with distilled water.

(2) Wait at least 30 s before taking the initial reading. Never adjust the volume of solution in a buret to exactly 0.00 mL. Attempting to do so will introduce bias into the measurement process and waste time.

(3) Now let about 5 mL of water run into a 250 mL Erlenmeyer flask. Wait at least 30 s

最终读数，计算每滴水的平均体积。重复上述操作，放出40滴水到锥形瓶中，再一次计算每滴水的平均体积。记录这两次结果并进行比较。

（5）练习滴加半滴水，计算半滴水的平均体积，比较半滴和一滴水的体积。在进行滴定时，应该尝试在半滴以内达到终点，以提高测量的精密度。

【数据记录与处理】

1. 定量转移

洗涤烧杯次数记录：

（1）_____次

（2）_____次

2. 移取溶液

首次洗涤蒸馏水用量 = _____ mL

二次洗涤蒸馏水用量 = _____ mL

三次洗涤蒸馏水用量 = _____ mL

结论：完全去除$KMnO_4$颜色所需最少洗涤次数为_____次，移液管最小清洗用水量为 _____ mL。

3. 滴定管的使用

（1）转移5 mL蒸馏水至锥形瓶。

初始读数 = _____ mL

最终读数 = _____ mL

转移体积 = _____ mL

（2）加30滴蒸馏水至锥形瓶。

初始读数 = _____ mL

最终读数 = _____ mL

一滴的平均体积 = _____ mL

（3）加40滴蒸馏水至锥形瓶。

初始读数 = _____ mL

最终读数 = _____ mL

一滴的平均体积= _____ mL

（4）加30次半滴蒸馏水至锥形瓶。

初始读数 = _____ mL

最终读数 = _____ mL

半滴的平均体积 = _____ mL

and take the final reading. The amount of solution in the Erlenmeyer flask is equal to the difference between the final reading and the initial reading. Record the final reading in your laboratory notebook, and then ask your instructor to take the final reading, too. Compare the two final readings. They should agree within 0.01 mL. Notice that the final digit in the buret reading is the estimate of the distance between two consecutive 0.1 mL marks on the buret.

(4) Refill the buret, and take a new zero reading. Now add 30 drops to the Erlenmeyer flask, and take the final reading. Calculate the mean volume of one drop; repeat this using 40 drops, and again calculate the mean volume of a drop. Record these results and compare them.

(5) Finally, practice adding half-drops to the flask. Calculate the mean volume of each half-drop, and compare your results with those that you obtained with full drops. When one performs titrations, one should attempt to determine endpoints to within half a drop to achieve good precision.

[Data Recording and Processing]

1. Making Quantitative Transfer

Number of Washing Beaker:

(1) _____

(2) _____

2. Delivering an Aliquot

Volume of distilled water for 1st washing =_____ mL

Volume of distilled water after 2nd washing =_____ mL

Volume of distilled water after 3rd washing =_____ mL

Thus: Minimum number of rinsings required to completely remove the color of $KMnO_4$ is

_____ ; minimum volume of distilled water used for washing pipet is_____ mL

3. Reading a Buret

(1) Transfer 5 mL of distilled water into the Erlenmeyer flask.

Initial reading =_____ mL

Final reading =_____ mL

Volume transferred =_____ mL

(2) Transfer 30 drops of distilled water into the Erlenmeyer flask.

Initial reading =_____ mL

Final reading =_____ mL

Mean of one drop =_____ mL

(3) Transfer 40 drops of distilled water into the Erlenmeyer flask.

Initial reading =_____ mL

Final reading =_____ mL

Mean of one drop =_____ mL

(4) Transfer 30 half-drops of distilled water into the Erlenmeyer flask.

Initial reading =_____ mL

Final reading =_____ mL

实验三　滴定分析操作练习

【实验目的】

(1) 熟练掌握滴定操作。

(2) 掌握酸碱标准溶液的配制方法、酸碱溶液相互滴定比较的方法。

(3) 熟悉甲基橙、酚酞指示剂的使用和滴定终点的正确判断，初步掌握酸碱指示剂的选择方法。

【实验原理】

滴定分析法是将滴定剂（已知准确浓度的标准溶液）滴加到含有被测组分的试液中，直到它们反应完全时为止，然后根据滴定剂的浓度和消耗的体积计算被测组分的含量的一种方法。在滴定分析实验中，必须学会标准溶液的配制、标定，滴定管的正确使用方法和滴定终点的正确判断方法。

浓 HCl 浓度不确定、易挥发，NaOH 不易制纯，在空气中易吸收 CO_2 和水分。因此，标准酸碱溶液要采用间接法配制，即先配制近似浓度的溶液，再用基准物质标定。

$0.1\ mol \cdot L^{-1}$ NaOH 溶液滴定相同浓度的 HCl 溶液时，滴定的突跃范围为 pH 4.3～9.7，可选用酚酞（变色范围 pH 8.0～9.6）和甲基橙（变色范围 pH 3.1～4.4）作指示剂。酚酞和甲基橙变色的可逆性好，当浓度一定的 NaOH 和 HCl 溶液相互滴定时，所消耗的体积比 $V(HCl)/V(NaOH)$ 应该是固定的。在使用同一指示剂的情况下，改变被滴定溶液的体积，此体积比应基本不变，借此，可训练学生的滴定基本操作技术和正确判断滴定终点的能力。通过观察滴定剂落点处周围颜色改变的快慢判断是否临近滴定终点；临近滴定终点时，要能控制一滴一滴地或半滴半滴地加入滴定剂，至最后一滴或半滴引起溶液颜色的明显变化，立即停止滴定，即为滴定终点。要做到这些，必须反复练习。

【仪器与试剂】

仪器：台秤、量筒（10 mL）、烧杯、试剂瓶、滴定管（50 mL）、锥形瓶（250 mL）、移液管（25 mL）。

Experiment 3   Practical Titration Exercise

【Objectives】

(1) To master the titration operation proficiently.

(2) To master the methods for preparing acid-base standard solutions and the comparison of inter-titration between acid and base solutions.

(3) To be familiar with the use of methyl orange and phenolphthalein indicators and the correct judgment of the endpoint. Master the selection of suitable acid and base indicators.

【Principle】

The titration is performed by slowly adding a titrant (a standard solution of known concentration) from a buret to a solution of the analyte until the reaction between the two is judged complete, and then the amount of the analyte is calculated based on the concentration and the volume consumed of the titrant. Therefore, in titrimetric experiment, one must learn the preparation and standardization of standard solutions, the correct use of burets and the correct judgment of the endpoint.

Concentrated HCl is volatile with uncertain concentration. NaOH is not easy to be purified since it is easy to absorb CO_2 and moisture in the air. Therefore, the acid and base standard solutions should be prepared by the indirect method, that is, solutions of approximate concentration are prepared first, and then standardized with primary standards.

When a 0.1 $mol \cdot L^{-1}$ NaOH solution titrates a HCl solution of equal concentration, the titration break range is approximately pH 4.3−9.7. Phenolphthalein (transition range pH 8.0 −9.6) and methyl orange (transition range pH 3.1 −4.4) can be used as indicators. Phenolphthalein and methyl orange have good reversibility of color change. When NaOH and HCl of a certain concentration are titrated with each other, the consumed volume ratio $V(HCl)/V(NaOH)$ should be fixed. If the same indicator is employed for different volumes of the analyte solution, this volume ratio should remain essentially unchanged. Through this, students' basic titration operation skills and the ability to correctly determine the endpoint can be trained. The approach to determine whether the endpoint is approaching is by observing the speed of color change in the vicinity of the titrant drop point. When the endpoint is approaching, it is necessary to be able to control the addition of the titrant drop by drop or half-drop by half-drop. When the last drop or half-drop causes a noticeable change in the

试剂：NaOH（固体）、浓 HCl（$\rho = 1.18$ g·mL^{-1}）、酚酞指示剂（0.2%乙醇溶液）、甲基橙指示剂（0.1%）、去离子水。

【实验步骤】

（一）酸碱标准溶液的配制

1. 0.1 mol·L^{-1} NaOH溶液的配制

用台秤迅速称取 2 g NaOH固体，放入100 mL小烧杯中，加约50 mL去离子水溶解，然后转移至试剂瓶中，用去离子水稀释至500 mL，摇匀后，用橡皮塞塞紧。贴好标签，写好试剂名称、浓度（空一格，留待填写准确浓度）、配制日期、班级、姓名等项。

2. 0.1 mol·L^{-1} HCl溶液的配制

用洁净量筒量取浓HCl约 4.5 mL，倒入 500 mL试剂瓶中，用去离子水稀释至500 mL，盖上玻璃塞，充分摇匀。贴好标签，备用。

（二）滴定操作练习

对碱式滴定管检漏，将其洗净，用 0.1 mol·L^{-1} NaOH溶液润洗2~3次（每次用5~10 mL溶液），然后装入 0.1 mol·L^{-1} NaOH溶液，排出气泡，调节滴定管液面至0.00 mL刻度。

对酸式滴定管检漏，将其洗净，用0.1 mol·L^{-1} HCl溶液润洗2~3次（每次用5~10 mL溶液），然后装入 0.1 mol·L^{-1} HCl溶液，排出气泡，调节滴定管液面至0.00 mL刻度。

从碱式滴定管中放出约20 mL NaOH溶液于250 mL锥形瓶中，再加1滴甲基橙指示剂，然后用酸式滴定管中的HCl溶液滴定锥形瓶中的NaOH溶液，进行滴定操作练习，同时观察指示剂颜色的变化。练习过程中，可在加入过量HCl溶液后再用NaOH溶液滴定HCl溶液，或在补加NaOH溶液后用HCl溶液滴定，如此反复或交替滴定，直至操作比较熟练后，再进行下面的实验。

（三）酸碱溶液的相互滴定

1. 用HCl溶液滴定NaOH（以甲基橙作指示剂）

从碱式滴定管中放出20~25 mL NaOH 溶液于250 mL锥形瓶中（注意：放出溶液时一般以每秒滴入3~4滴溶液为宜，若溶液放出速度较快，则应稍等一下后再读数），再加入1~2滴甲基橙指示剂，在不断摇动下，用HCl溶液滴定至溶液由黄色变为橙色即为终点，记下读数。如此操作，再滴定2份。计算体积比 $V(\text{HCl})/V(\text{NaOH})$，要求相对偏差在 ±0.3%以内。

color of the solution, the titration is stopped immediately, which is the endpoint. To achieve this, repeated practice is essential.

【Apparatus and Chemicals】

Apparatus: scale, measuring cylinder (10 mL), beaker, reagent bottle, buret (50 mL), Erlenmeyer flask (250 mL), pipet (25 mL).

Chemicals: NaOH (solid), concentrated HCl ($\rho = 1.18$ g·mL^{-1}), phenolphthalein indicator (0.2% ethanol solution), methyl orange indicator (0.1%), deionized water.

【Procedure】

I. Preparation of Acid-Base Standard Solutions

1. Preparation of 0.1 mol·L^{-1} NaOH solution

Rapidly weigh 2 g of NaOH solid in a 100 mL beaker. Add approximately 50 mL of deionized water to dissolve, then transfer it to a reagent bottle. Dilute it to 500 mL with deionized water, shake well, and seal it tightly with a rubber stopper. Label the bottle with the name of reagent, concentration (leave a space for filling in the exact concentration), preparation date, class, student name, etc.

2. Preparation of 0.1 mol·L^{-1} HCl solution

Add 4.5 mL of concentrated HCl into a 500 mL reagent bottle with a clean measuring cylinder. Dilute it to 500 mL with deionized water, cover it with a glass stopper, and shake thoroughly. Affix a label and set it aside for use.

II. Titration Practice

Retrieve a base buret. The buret is checked leakage and cleaned, and then rinsed 2–3 times with the 0.1 mol·L^{-1} NaOH solution (5–10 mL each time). Fill the buret with the 0.1 mol·L^{-1} NaOH solution. Free of air bubbles inside the tip and adjust the liquid level of the buret to the 0.00 mL mark.

Retrieve an acid buret. The buret is tested for leakage and cleaned, and then rinsed 2–3 times with the 0.1 mol·L^{-1} HCl solution (5–10 mL each time). Fill the buret with the 0.1 mol·L^{-1} HCl solution. Remove bubbles and adjust the liquid level to 0.00 mL.

About 20 mL of the NaOH solution is released from the base buret into a 250 mL Erlenmeyer flask, and 1 drop of methyl orange indicator is added. Then, the NaOH solution in the Erlenmeyer flask is titrated with the HCl solution from the acid buret. Titration operation is practiced, and the color change of the indicator is observed. During the exercise, the HCl solution can be titrated with NaOH solution after adding excessive HCl solution, or titrated with HCl solution after adding NaOH solution. Do this repeatedly or alternately until being skillful before conducting the following experiment.

III. Inter-titration of Acid-Base Solutions

1. Titration of NaOH with HCl using methyl orange as the indicator

Deliver 20–25 mL of NaOH from the base buret into a 250 mL Erlenmeyer flask (Note:

2. 用NaOH溶液滴定HCl（以酚酞作指示剂）

用移液管吸取25 mL HCl溶液于250 mL锥形瓶中，加入2～3滴酚酞指示剂，在不断摇动下，用 NaOH 溶液滴定，注意控制滴定速度，当滴加的NaOH溶液落点处周围红色褪去较慢时，表明临近滴定终点，用洗瓶洗涤锥形瓶内壁，控制NaOH溶液一滴一滴地或半滴半滴地滴出。至溶液呈微红色，且半分钟不褪色即为滴定终点，记下读数。如此平行滴定3份，要求各次所消耗NaOH溶液体积的最大差值不超过 ± 0.04 mL。

【数据记录与处理】

把数据记录在表3.3、表3.4中。

表3.3   HCl滴定NaOH（以甲基橙作指示剂）

项目	序号		
	I	II	III
$V(NaOH)/mL$			
$V(HCl)/mL$			
$V(HCl)/V(NaOH)$			
$V(HCl)/V(NaOH)$平均值			
$RAD/\%$			

表3.4   NaOH滴定HCl（以酚酞作指示剂）

项目	序号		
	I	II	III
$V(HCl)/mL$	25.00	25.00	25.00
$V(NaOH)/mL$			
$V(NaOH)$平均值/mL			
$V(NaOH)$的极差/mL			

【思考题】

（1）HCl和NaOH标准溶液能否用直接法配制？为什么？

（2）配制NaOH溶液时，应选用何种天平称取试剂？为什么？

（3）将标准溶液装入滴定管之前，为什么要用该溶液润洗滴定管2～3次？而锥形瓶是否也需用该溶液润洗后烘干？为什么？

It is advisable to add 3 to 4 drops per second when releasing the solution. If the solution is released too quickly, wait for a moment before taking the reading). Add 1–2 drops of methyl orange indicator. Under continuous swirling, titrate with HCl solution until the yellow color turns exactly orange as the endpoint, and record the reading. Repeat this operation twice more. Calculate the volume ratio of $V(HCl)/V(NaOH)$, and the relative deviation is required to be within ± 0.3%.

2. Titration of HCl with NaOH using phenolphthalein as the indicator

Pipet 25 mL of HCl solution into a 250 mL Erlenmeyer flask. Add 2–3 drops of phenolphthalein indicator, and titrate with NaOH solution under constant swirling. Pay attention to controlling the titration rate. When the red color fades slowly in the vicinity of the drop point, it indicates that the endpoint is approaching. Rinse the inside of the Erlenmeyer flask with a wash bottle and add the NaOH solution drop by drop or half-drop by half-drop. Note the reading until the solution turns slightly pink and does not fade within half a minute. Titrate three replicates in this way; the maximum difference in the volume of NaOH solution consumed in each titration should not exceed ± 0.04 mL.

【Data Recording and Processing】

Record the experimental data in Table 3. 3 and Table 3. 4.

Table 3. 3   Titration of NaOH with HCl （Indicator： Methyl Orange）

Item	No.		
	I	II	III
$V(NaOH)$/mL			
$V(HCl)$/mL			
$V(HCl)/V(NaOH)$			
Average of $V(HCl)/V(NaOH)$			
RAD/%			

Table 3. 4   Titration of HCl with NaOH （Indicator： Phenolphthalein）

Item	No.		
	I	II	III
$V(HCl)$/mL	25.00	25.00	25.00
$V(NaOH)$/mL			
Average of $V(NaOH)$/mL			
Range of $V(NaOH)$/mL			

【Questions】

(1) Can HCl and NaOH standard solutions be prepared by the direct method? Why?

(2) When preparing the NaOH solution, which type of balance should be selected to

（4）滴定管、移液管量取溶液体积时，读数时应记录几位有效数字？

（5）滴定管读数的起点为何每次均要调到零刻度附近？

（6）为什么用HCl溶液滴定NaOH溶液时一般采用甲基橙作为指示剂，而用NaOH溶液滴定HCl溶液时以酚酞作为指示剂？

weigh the reagents? Why?

(3) Why should the buret be rinsed 2 or 3 times with the standard solution before being filled? And is it necessary to rinse or dry the Erlenmeyer flask with the solution? Why?

(4) When measuring the volume of solutions with burets or pipets, how many significant digits should be recorded?

(5) Why is it necessary to adjust the starting point of the buret reading to near the "0.00" mark every time?

(6) Why is methyl orange typically employed as an indicator in the titration of NaOH solution with HCl solution, whereas phenolphthalein is preferred when titrating HCl solution with NaOH solution?

实验四　容量器皿的校准

【实验目的】

（1）掌握滴定管、移液管、容量瓶的使用方法。

（2）了解容量器皿校准的意义，并学习滴定管、移液管、容量瓶的校准方法。

（3）学习移液管和容量瓶之间的相对校准。

【实验原理】

滴定管、移液管和容量瓶是分析化学实验中所用的主要量器，用于量取准确体积的溶液。这些容量器皿由制造商标记其校准的温度，通常为20 ℃。实际上，我们经常在其他温度下使用容量器皿，导致容量器皿的容积与其刻度并非完全相符合。因此，在准确度要求较高的分析工作中，必须对容量器皿进行校准。

容量器皿的校准常采用两种方法。

1. 称量法

称量法是通过测量容器所量入或量出的已知密度和温度的液体（通常是蒸馏水或去离子水）的质量来实现的。

校准容量器皿时，应考虑以下因素：

（1）浮力引起的质量变化。

（2）温度变化引起液体密度变化。

（3）温度变化引起玻璃容器体积变化。

利用表3.5数据可以进行体积校准，表中的换算系数包含了对不锈钢或黄铜砝码（两者之间的密度差小到可以忽略）的浮力修正以及对水和玻璃容器的体（容）积变化的修正。将T温度下的水质量乘以表3.5中相应的换算系数，可得到水在该温度下相应的体积或20℃时的体积。校准相关的计算如例1所示。

Experiment 4　Calibration of Volumetric Glassware

【Objectives】

(1) To master the operation of burets, pipets and volumetric flasks.

(2) To comprehend the significance of glassware calibration and study the calibration approaches of burets, pipets, and volumetric flasks.

(3) To learn the relative calibration between a pipet and a volumetric flask.

【Principles】

In analytical chemistry experiment, volume may be measured reliably with a buret, a pipet, or a volumetric flask. Volumetric equipment is marked by the manufacturer to indicate the temperature at which the calibration strictly applies. This temperature is normally 20℃. In practice, we frequently use the volumetric glassware in other temperature, resulting in bias between the actual volume and marked volume of glassware. So for experiments requiring highly precise results, it's necessary to calibrate the volumetric equipments.

Generally, the following two methods are employed in the calibration of volumetric glassware:

1. Weighing

By the method, volumetric glassware is calibrated by measuring the mass of liquid (usually distilled or deionized water) of known density and temperature that is contained in (or delivered by) the volumetric glassware.

To calibrate volumetric equipment, the following factors should be considered.

(1) Mass change by buoyancy.

(2) Density change of liquid with temperature.

(3) Volume change of glass containers with temperature.

Table 3.5 is provided to help with buoyancy calculations. Corrections for buoyancy with respect to stainless steel or brass mass (the density difference between the two is small enough to be neglected) and for the volume change of water and of glass containers have been incorporated into these data. Multiplication by the appropriate factor from Table 3.5 converts the mass of water at temperature T to the corresponding volume at that temperature or the volume at 20℃. The related calculation is shown in Example 1.

例1 用25 mL移液管移取24.976 g水（25 ℃下，用不锈钢砝码称重），请用表3.5中的数据计算出该移液管在25 ℃和20 ℃时的容积。

25 ℃时：V=24.976 g×1.004 0 mL/g=25.08 mL

20 ℃时：V=24.976 g×1.003 7 mL/g=25.07 mL

表3.5 在空气中用不锈钢砝码称量的1.000 g水所占的体积

温度T/℃	体积/mL	
	T 时	校正到20 ℃
10	1.001 3	1.001 6
11	1.001 4	1.001 6
12	1.001 5	1.001 7
13	1.001 6	1.001 8
14	1.001 8	1.001 9
15	1.001 9	1.002 0
16	1.002 1	1.002 2
17	1.002 2	1.002 3
18	1.002 4	1.002 5
19	1.002 6	1.002 6
20	1.002 8	1.002 8
21	1.003 0	1.003 0
22	1.003 3	1.003 2
23	1.003 5	1.003 4
24	1.003 7	1.003 6
25	1.004 0	1.003 7
26	1.004 3	1.004 1
27	1.004 5	1.004 3
28	1.004 8	1.004 6
29	1.005 1	1.004 8
30	1.005 4	1.005 2

2. 相对校准法

用一个已校准的玻璃容器间接地校准另一个玻璃容器，即相对校准法。在滴定分析中，要求两种容器体积之间有一定的比例关系时，常采用此法。例如，25 mL移液管量取液体的体积应等于250 mL容量瓶量取体积的十分之一。

Example 1

A 25 mL pipet delivers 24.976 g of water weighed against stainless steel mass at 25 ℃. Use the data in Table 3.5 to calculate the volume delivered by this pipet at 25 ℃ and 20 ℃.

At 25 ℃: $V = 24.976 \text{ g} \times 1.004\,0 \text{ mL/g} = 25.08 \text{ mL}$

At 20 ℃: $V = 24.976 \text{ g} \times 1.003\,7 \text{ mL/g} = 25.07 \text{ mL}$

Table 3. 5   Volume Occupied by 1. 000 g of Water Weighed in Air against Stainless Steel Weights

Temperature T/℃	Volume/mL	
	At T	Corrected to 20 ℃
10	1. 001 3	1. 001 6
11	1. 001 4	1. 001 6
12	1. 001 5	1. 001 7
13	1. 001 6	1. 001 8
14	1. 001 8	1. 001 9
15	1. 001 9	1. 002 0
16	1. 002 1	1. 002 2
17	1. 002 2	1. 002 3
18	1. 002 4	1. 002 5
19	1. 002 6	1. 002 6
20	1. 002 8	1. 002 8
21	1. 003 0	1. 003 0
22	1. 003 3	1. 003 2
23	1. 003 5	1. 003 4
24	1. 003 7	1. 003 6
25	1. 004 0	1. 003 7
26	1. 004 3	1. 004 1
27	1. 004 5	1. 004 3
28	1. 004 8	1. 004 6
29	1. 005 1	1. 004 8
30	1. 005 4	1. 005 2

2. Relative calibration

In relative calibration, one calibrated glassware can be used to indirectly calibrate another. This method is often used in titrimetric analysis when a certain proportion relationship between the volumes of two vessels is required. For example, the volume of liquid delivered from a 25 mL pipet is one-tenth aliquot of that in a 250 mL volumetric flask.

【仪器与试剂】

分析天平、滴定管（50 mL）、移液管（25 mL）、容量瓶（250 mL）、锥形瓶（125 mL，带磨口玻璃塞）、温度计（0 ℃～50 ℃或0 ℃～100 ℃）、洗耳球。

【实验步骤】

1. 滴定管的校准

（1）在一个干净的滴定管中加入蒸馏水，排除管尖气泡。等待约1 min后，降低液面至零刻度处。如滴定管尖有液体，应将其触碰烧杯内壁除去。10 min后，重新读数，如果读数不变，表明滴定管旋塞密封性好，不漏水。在此期间进行步骤（2）和（3）。

（2）用分析天平称一洗净且外表干燥的带磨口玻璃塞的125 mL锥形瓶重量，记录至小数点后第三位（0.001 g）。称重后请勿用手指触摸锥形瓶，可用夹子或折叠的纸条来取锥形瓶。

（3）测量并记录水温。

（4）检查滴定管旋塞密封性后，缓慢地（约10 mL/min）将约10 mL的水注入锥形瓶中，将管尖触碰锥形瓶内壁。等待1 min，记录放出水的体积（V_0）。称量锥形瓶和水的总质量，记录至0.001 g。这个质量和空瓶质量之差就是所放出水的质量。使用表3.5中数据将这个质量转换为实际容量V_{20}。

（5）用此法称量每次从滴定管中放出的约5 mL或10 mL水（记为V_0）的质量，直到放至50 mL。用换算系数表（见表3.5）乘每次得到的水的质量，即可得到滴定管各部分的实际容量V_{20}。重复校准一次，两次相应区间纯水的质量相差应小于0.02 g，求出平均值，并计算校准值$\Delta V = V_{20} - V_0$。

（6）表3.6为在水温25 ℃校准的一支50 mL滴定管的部分实验数据。最后一项为总校准值。校准时也可每次都从滴定管的零刻度或稍低处开始分别放出不同体积（如10 mL、20 mL、30 mL）的纯水后称量，求得总校准值。

（7）移液管和吸量管也可采用上述称量法进行校准。用称量法校准容量瓶时，不必用锥形瓶称量，且校准至0.01 g即可。

【Apparatus and Chemicals】

Analytical balance, buret (50 mL), pipet (25 mL), volumetric flask (250 mL), Erlenmeyer flask (125 mL, with ground glass stopper), thermometer (0 ℃–50 ℃ or 0 ℃–100 ℃), rubber suction bulb.

【Procedure】

1. Calibration of a buret

(1) Fill a clean buret with distilled water and make sure that no air bubbles are trapped in the tip. Allow about 1 min for drainage; then lower the liquid level to bring the bottom of the meniscus to the 0.00 mL mark. Touch the tip to the wall of a beaker to remove any adhering drop. Wait 10 min and recheck the volume; if the stopper is tight, there should be no perceptible change. During this interval, do steps (2) and (3).

(2) Weigh (to the nearest 0.001 g) a 125 mL Erlenmeyer flask fitted with a ground glass stopper. Do not touch the flask with your fingers after this weighing. Use tongs or a folded strip of waxed paper to manipulate the flask.

(3) Measure and record the temperature of the water.

(4) Once tightness of the stopper has been established, slowly transfer (at about 10 mL/min) approximately 10 mL of water to the flask. Touch the tip to the wall of the flask. Wait 1 min, record the volume (V_0) that was apparently delivered. Weigh the flask and its contents to the nearest 0.001 g. The difference between this mass and the initial value gives the mass of water delivered. Use Table 3. 5 to convert this mass to the true volume (V_{20}).

(5) Repeat the calibration, test the buret at 5 mL or 10 mL (marked as V_0) intervals over its entire volume. Calculate the actual volume (V_{20}) of each part of the buret by multiplying the conversion factor at the experimental temperature (see Table 3.5) by the mass of water obtained each time. Repeat the calibration until agreement within ±0.02 g is achieved. Calculate the average and the correction value ($\Delta V = V_{20} - V_0$).

(6) Table 3.6 shows partial experimental data of calibrating a 50 mL buret at 25 ℃. The last term in the table is the overall correction value. The buret can also be calibrated by adding different volumes of water (such as 10 mL, 20 mL and 30 mL) from the 0.00 mL mark (or slightly lower) each time to obtain the overall correction.

(7) Pipets can also be calibrated using the above weighing method. When using the weighing method to calibrate a volumetric flask, it is not necessary to use an Erlenmeyer flask, and weigh to the nearest 0.01 g.

表 3.6　50 mL滴定管校准表

V_0 / mL	瓶和水的质量/g	瓶的质量/g	水的质量/g	V_{20} / mL	ΔV / mL
0.00~10.10	39.280		10.080	10.117	+0.02
0.00~20.07	49.190		19.990	20.064	−0.01
0.00~30.05	59.180	29.200	29.980	30.091	+0.04
0.00~40.00	69.130		39.930	40.078	+0.08
0.00~49.94	79.010		49.810	49.994	+0.05

注：①水的温度为25 ℃，换算为20 ℃体积的系数为1.003 7；②V_0：从滴定管放出的水的体积；③V_{20}：换算为20 ℃时的真实体积；④$\Delta V = V_{20} - V_0$。

2. 容量瓶与移液管的相对校准

用洁净的25 mL移液管吸取去离子水，注入洁净、干燥的250 mL容量瓶中（操作时切勿让水碰到容量瓶的磨口）。重复10次，然后观察溶液弯液面下缘是否与刻度线相切，若不相切，另做新标记，经相互校准后的容量瓶与移液管均做上相同记号，可配套使用。

【注意事项】

（1）在校准之前，所有的容量器皿都应该彻底清洗干净，使内壁不挂水珠。滴定管和移液管无须干燥，容量瓶应于室温下晾干。

（2）用于校准的水应与其周围环境温度一致。为了确保这一条件，需要提前取好水，每隔一段时间就检测水温，直到水温不变为止。

【思考题】

（1）称量水的质量时，为什么只要精确至0.001 g?

（2）为什么要进行容量器皿的校准? 影响容量器皿体积刻度使之不准确的主要因素有哪些?

（3）从滴定管放蒸馏水到称量的容量瓶内时，应注意什么?

（4）校准容量瓶时为什么需要晾干? 在用容量瓶配制标准溶液时是否也要晾干?

Table 3. 6　Data Sheet for 50 mL Buret Calibration

V_0 / mL	Mass of flask and water/g	Mass of flask/g	Mass of water/g	V_{20} / mL	ΔV / mL
0.00–10.10	39.280		10.080	10.117	+0.02
0.00–20.07	49.190		19.990	20.064	−0.01
0.00–30.05	59.180	29.200	29.980	30.091	+0.04
0.00–40.00	69.130		39.930	40.078	+0.08
0.00–49.94	79.010		49.810	49.994	+0.05

Note： ① Water temperature： 25 ℃ , conversion factor to 20 ℃ : 1.003 7; ② V_0 : Volume of water delivered; ③ V_{20} : Volume at 20 ℃ ; ④ $\Delta V = V_{20} - V_0$.

2.　Calibrating a volumetric flask relative to a pipet

Carefully transfer ten times of 25 mL aliquots of deionized water with a clean pipet in to a clean and dry 250 mL volumetric flask (do not let the water touch the ground spout of the volumetric flask). Observe whether the bottom of the meniscus coincide with the graduation mark. If not, mark the location of the meniscus. Make the same mark on the volumetric flask and pipet after combination of volumes, and they can be used together.

【Notes】

(1)　All volumetric glassware should be painstakingly free of water breaks before being calibrated. Burets and pipets need not be dry; volumetric flasks should be thoroughly drained and dried at room temperature.

(2)　The water used for calibration should be in thermal equilibrium with its surroundings. This condition is best established by drawing the water well in advance, noting its temperature at frequent intervals, and waiting until no further changes occur.

【Questions】

(1)　In the calibration of a buret, why is it read to the nearest 0.001 g when weighing?

(2)　Why should volumetric equipment be calibrated? What factors should be considered in the calibration?

(3)　What should be paid attention to when delivering distilled water from the buret into the volumetric flask to be weighed?

(4)　Why must volumetric flasks be dried before calibration? Should they be dried when preparing standard solutions?

第四章　酸碱滴定实验

Chapter 4　Acid-Base Titration

实验五　氢氧化钠标准溶液的标定

【实验目的】

（1）掌握滴定操作技术。

（2）掌握用邻苯二甲酸氢钾（$KHC_8H_4O_4$）作为基准物标定氢氧化钠（NaOH）溶液的准确浓度的方法。

（3）掌握NaOH标准溶液的配制方法。

【实验原理】

一未知浓度的溶液可通过滴加已知浓度的第二种溶液（基准物）与之反应直到刚好反应完全，来对其进行定量分析，这就是滴定分析法。在滴定过程中，需要用一个滴定管来滴加滴定剂到装有被分析物的锥形瓶。滴定剂可以是已知或未知浓度的溶液，被分析物可以是用移液管准确移取的一定体积的溶液，也可以是质量经准确称量的溶解固体。当两种反应物质的摩尔比与配平后的方程式相同时，反应即完成，此时为滴定中的化学计量点或等当点。本实验用酚酞指示剂测定酸碱滴定的等当点。酚酞在酸性溶液中无色，在碱性溶液中为红色，酚酞指示剂变色的点即为滴定终点，滴定终点与等当点越接近越好。

NaOH具有吸湿性，无法直接准确以之配制标准溶液，需要用一种酸作为基准物来确定其准确浓度。在本实验中，用干燥的纯$KHC_8H_4O_4$作为测定NaOH标准溶液准确浓度的基准物质。$KHC_8H_4O_4$由于纯度高、摩尔质量较大和不吸湿而具有基准物的性质。$KHC_8H_4O_4$的物质的量由其质量和摩尔质量（204.44 $g \cdot mol^{-1}$）计算得出：

$$n(KHC_8H_4O_4) = \frac{m(KHC_8H_4O_4)}{M(KHC_8H_4O_4)}$$

$$= \frac{m(KHC_8H_4O_4)}{204.44 \ g \cdot mol^{-1}}$$

Experiment 5　Standardization of NaOH Solution

【Objectives】

(1) To master the operation technique of titration.

(2) To master the accurate determination of the concentration of sodium hydroxide (NaOH) solution by using potassium hydrogen phthalate (KHC$_8$H$_4$O$_4$) as the primary standard.

(3) To master the preparation method of the standard NaOH solution.

【Principle】

A solution of unknown concentration can be analyzed quantitatively by adding a second solution of known concentration (primary standard solution) until the reaction is complete. This procedure is known as titration. The titration procedure requires a buret to dispense a liquid, called the titrant, into a flask containing the analyte. The titrant may be a solution of known or unknown concentration. The analyte may be a solution whose volume is measured with a pipet or it may be a dissolved solid with a very accurately measured mass. The reaction is complete when the molar ratio of the two reacting substances is the same as what appears in the balanced equation. This is the stoichiometric point or equivalence point in the titration. In this experiment, the equivalence point for an acid-base titration is detected using phenolphthalein indicator, colorless in acidic and red in basic solutions. The point at which the phenolphthalein indicator changes color is the endpoint of the titration. Indicators are selected so that the equivalence point and the endpoint occur at essentially the same point in the titration.

Solutions of NaOH are virtually impossible to prepare to a precise molar concentration because the substance is hygroscopic. To prepare a NaOH solution with an exact molar concentration, it must be standardized with an acid that is a primary standard. In this experiment, dry KHC$_8$H$_4$O$_4$ is used as the primary acid standard for determining the molar concentration of NaOH solution. KHC$_8$H$_4$O$_4$ has the properties of a primary standard because of its high purity, relatively high molar mass, and because it is not hygroscopic. The amount of KHC$_8$H$_4$O$_4$ used for the analysis is calculated from its measured mass and molar mass (204.44 g·mol^{-1}):

$$n(\mathrm{KHC_8H_4O_4}) = \frac{m(\mathrm{KHC_8H_4O_4})}{M(\mathrm{KHC_8H_4O_4})}$$

$$= \frac{m(\mathrm{KHC_8H_4O_4})}{204.44 \text{ g·mol}^{-1}}$$

根据反应式：$HC_8H_4O_4^-(aq)+OH^-(aq)\longrightarrow H_2O(l)+C_8H_4O_4^{2-}(aq)$，1 mol $KHC_8H_4O_4$ 与 1 mol NaOH 刚好反应完全。

实验中，将准确称量的 $KHC_8H_4O_4$ 基准物溶解于去离子水中，然后将配制好的 NaOH 溶液从滴定管中滴入 $KHC_8H_4O_4$ 溶液，直到酚酞指示剂由无色变为粉红色且保持30 s不变，记录加入的 NaOH 溶液体积。根据下式计算 NaOH 溶液的摩尔浓度：

$$c(NaOH)=\frac{n(NaOH)}{V(NaOH)}$$

$$=\frac{m(KHC_8H_4O_4)}{204.44\times V(NaOH)}$$

一旦计算出 NaOH 溶液的摩尔浓度，溶液就称为"已标定"，此 NaOH 溶液称为二级标准溶液。

【仪器与试剂】

仪器：滴定管、滴定管夹、铁架台、锥形瓶（250 mL×3）、称量瓶、移液管（25 mL）。

试剂：NaOH（固体）、$KHC_8H_4O_4$、酚酞指示剂。

【实验步骤】

（1）制备 0.1 mol·L^{-1} NaOH 溶液。

在实验前一周，溶解约4 g NaOH固体于少量蒸馏水中，转移到一个带橡胶塞的瓶子里，并稀释到1 L。

注意：NaOH溶液具有腐蚀性，避免洒在皮肤或衣服上。

（2）制备基准物 $KHC_8H_4O_4$ 标准溶液。

取3个洁净的锥形瓶（编号1~3），以减量法准确称量3份0.4~0.6 g的 $KHC_8H_4O_4$ 基准物（将基准物敲出置于锥形瓶中），记录每份基准物的准确质量（读至0.1 mg）。加入 40~50 mL的蒸馏水，摇匀溶解（必要时，可以加热溶解固体酸）。然后在每个锥形瓶中各加入2滴酚酞。

（3）准备滴定管和滴定剂。

润洗滴定管：清洗一个50 mL的滴定管（干净的滴定管应没有液滴附着在内壁）。用蒸馏水润洗三次，再用NaOH溶液润洗三次（确保润湿整个内表面，一定要让一些液体从滴定管尖端流出）。

装入NaOH溶液：每次装入NaOH溶液后，从滴定管尖端放出约2 mL溶液，以确保尖端干净且NaOH已经充满尖端（没有气泡）。

将滴定管固定在铁架台的夹子上。

From the balanced equation for the reaction, 1 mole of $KHC_8H_4O_4$ reacts with 1 mole of NaOH according to the net ionic equation:

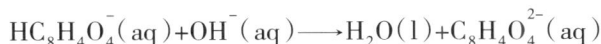

$$HC_8H_4O_4^-(aq)+OH^-(aq)\longrightarrow H_2O(l)+C_8H_4O_4^{2-}(aq)$$

In the experimental procedure, an accurately measured mass of dry $KHC_8H_4O_4$ is dissolved in deionized water. A prepared NaOH solution is then dispensed from a buret into the $KHC_8H_4O_4$ solution until the stoichiometric point is reached, signaled by the colorless to pink change of the phenolphthalein indicator. At this point the dispensed volume of NaOH is noted and recorded.

The molar concentration of the NaOH solution is calculated using the following equation:

$$c(NaOH)=\frac{n(NaOH)}{V(NaOH)}$$

$$=\frac{m(KHC_8H_4O_4)}{204.44 \times V(NaOH)}$$

Once the molar concentration of the NaOH is calculated, the solution is said to be "standardized" and the NaOH solution is called a secondary standard solution.

【Apparatus and Chemicals】

Apparatus: buret, buret clamp, ring stand, Erlenmeyer flask (250 mL × 3), weighing bottle, pipet (25 mL).

Chemicals: NaOH (solid), $KHC_8H_4O_4$, phenolphthalein indicator.

【Procedure】

(1) Preparing $0.1\ mol \cdot L^{-1}$ NaOH solution.

One week before the scheduled laboratory period, dissolve about 4 g of NaOH (solid) in the minimum amount of distilled water, transfer to a rubber-stoppered bottle and dilute it to 1 L.

Note: NaOH solution is caustic. Avoid spilling it on your skin or clothing.

(2) Making the primary standard acid solution: $KHC_8H_4O_4$.

Label 3 Erlenmeyer flasks (1–3). Weigh out (by difference) three samples of $KHC_8H_4O_4$, 0.4–0.6 g each. Make sure to write down the mass of each (1–3) on data sheet. Dissolve each with 40–50 mL of distilled water (If necessary, you may warm the solution to dissolve all the solid acid). Add 2 drops of phenolphthalein to each flask.

(3) Preparing the buret and titrant.

Clean a 50 mL buret (The buret is clean enough when water droplets do not cling to the inner surface). Rinse with distilled water three times, and then rinse with three portions of NaOH solution. Make sure to wet the entire inner surface and let some out the buret tip. Fill your buret with NaOH solution. Always dispense about 2 mL to make sure the tip is clean and NaOH has filled the tip (no air bubbles).

Place the buret in clamp which is attached to ring stand.

(4) Titration.

Record the initial buret volume (V_0) on data sheet, read to 0.01 mL. Place a piece of white paper under trial 1 of the $KHC_8H_4O_4$ in No. 1 Erlenmeyer flask which should now be under

（4）滴定。

读取滴定管初始体积V_0（读到0.01 mL）后，将锥形瓶1（含第一份$KHC_8H_4O_4$基准物和指示剂）置于滴定管下，并在锥形瓶下放置一张白纸。慢慢将NaOH溶液滴加到$KHC_8H_4O_4$溶液中，边滴边摇动锥形瓶。随着NaOH溶液的加入，在滴入点附近产生红色，摇动锥形瓶，红色消失，当红色消失的速率变慢时，滴加速率也随之放慢；用洗瓶中的蒸馏水冲洗锥形瓶内部，继续滴加NaOH溶液。在终点附近，加入半滴NaOH溶液，用力摇动锥形瓶，如果溶液变为淡粉色且持续至少30 s，即为终点。在实验表上记录滴定管最终体积V_1（读到0.01 mL），计算所用的NaOH溶液体积V_1-V_0（mL）。

重新装满滴定管，重复前述步骤，滴定另两份质量不同但已知准确质量的$KHC_8H_4O_4$。

（5）计算NaOH溶液的摩尔浓度$c(NaOH)$和相对平均偏差RAD，在装有NaOH溶液的玻璃瓶上标上实验人员姓名和该溶液的摩尔浓度。

【数据记录与处理】

将数据记录在表4.1中。

表4.1　NaOH溶液的制备及标定

项目	序号		
	I	II	III
$m(KHC_8H_4O_4)/g$			
V_0/mL			
V_1/mL			
V_1-V_0/mL			
$c(NaOH)/(mol \cdot L^{-1})$			
$\bar{c}(NaOH)/(mol \cdot L^{-1})$			
$RAD/\%$			

【注意事项】

（1）在烧杯中称NaOH固体，不要在纸上称。

（2）在每个试剂瓶上贴上标签，注明试剂名称、配制日期、实验人员姓名、试剂浓度等。

（3）在加入NaOH溶液之前，用NaOH溶液润洗滴定管三次。

（4）务必去除滴定管尖端的气泡。

the NaOH filled buret. Slowly add the NaOH solution to the $KHC_8H_4O_4$, swirling the flask after eachaddition. As the rate of the indicator color change decreases, decrease the rate of NaOH addition; rinse the wall of the flask with distilled water from your wash bottle, then proceed with drop addition of NaOH. In the immediate vicinity of the endpoint, add a half drop until the indicator endpoint is reached. This is when a half drop causes the faint pink color to last for at least 30 s. Record the final buret volume (V_1), read to 0.01 mL. Calculate the volume of NaOH solution dispensed and record.

Refill the buret and repeat the titration at least two more times with varying, but accurately known, masses of $KHC_8H_4O_4$.

(5) Calculate the molar concentration of the NaOH solution and the relative average deviation (RAD). Label the glass bottle containing the NaOH solution with the student's name and the content's molar concentration.

【Data Recording and Processing】

Record the experimental data in Table 4.1.

Table 4.1　Preparation and Standardization of NaOH Solution

Item	No.		
	I	II	III
$m(KHC_8H_4O_4)/g$			
V_0/mL			
V_1/mL			
V_1-V_0/mL			
$c(NaOH)/(mol \cdot L^{-1})$			
$\bar{c}(NaOH)/(mol \cdot L^{-1})$			
$RAD/\%$			

【Notes】

(1) Weigh NaOH (solid) in a beaker instead of on a piece of paper.

(2) Put a label on each reagent bottle with the name of the reagent, date of preparation, operator, concentration, etc.

(3) Rinse the buret with NaOH solution three times before filling it with the NaOH solution.

(4) Get rid of air bubbles from the tip of the buret if there is any.

(5) Adjust liquid level to near zero point before each titration.

(6) Don't overshoot your endpoint!

(7) Accuracy and precision will be used in grading.

(8) Caustic standard solutions corrode glass, long-term storage is best in plastic bottles. Generally, they can be stored in glass bottles, but rubber plugs must be used. The prepared NaOH solution should be tightly stoppered and stored in the drawer.

（5）每次滴定前将液面调整到接近零刻度。

（6）不要超过滴定终点。

（7）根据实验结果的准确度和精密度进行评分。

（8）苛性碱标准溶液会侵蚀玻璃，长期保存最好用塑料瓶，在一般情况下，可用玻璃瓶贮存，但必须用橡皮塞。配好的NaOH溶液要盖紧瓶塞，存储于试剂柜中。

（9）离开实验室前请指导教师检查实验数据并签名。

【思考题】

（1）写出实验的计算过程。

（2）说明如何从滴定管中滴出半滴溶液。

（3）用纯$KHC_8H_4O_4$标定NaOH标准溶液，假设$KHC_8H_4O_4$没有完全干燥，则计算得到的NaOH溶液的摩尔浓度是过高、过低，还是不变？为什么？

（4）实验过程建议加入2滴酚酞用于指示滴定终点，请解释为什么不能加入太多指示剂（比如20滴酚酞）。

（5）在NaOH溶液的标定过程中，如果滴定开始前滴定管尖端有一个气泡，在滴定过程中，气泡从滴定管流出，那么所得的NaOH溶液的摩尔浓度是过高、过低还是不变？为什么？

（6）本实验中，$KHC_8H_4O_4$的质量必须准确称量到0.1 mg，然而溶解它的水的体积却无须准确量取，甚至在滴定过程中还用水清洗锥形瓶内壁。请解释为什么加水到锥形瓶中对分析结果影响不大，但如果加水到NaOH溶液中则有很大影响。

(9) Get the instructor's sign off before leaving.

[Questions]

(1) Show calculation for the experiment.

(2) Explain how a half-drop of titrant can be dispensed from a buret.

(3) Pure $KHC_8H_4O_4$ is used for the standardization of the NaOH solution. Suppose that the $KHC_8H_4O_4$ is not completely dry. Will the reported molar concentration of the NaOH solution be too high, too low, or unaffected because of the moistness of the $KHC_8H_4O_4$? Explain why.

(4) The experimental procedure suggests the addition of 2 drops of phenolphthalein for the standardization of the NaOH solution. Explain why the analysis will be less accurate with the addition of a larger amount (e.g., 20 drops of phenolphthalein).

(5) Suppose an air bubble is initially entrapped in the buret for the standardization of the NaOH solution. However, during the titration the air bubble is passed from the buret. Will the reported molar concentration of the NaOH be calculated as too high, too low, or unchanged as a result? Explain why.

(6) The mass of $KHC_8H_4O_4$ is measured to the nearest 0.1 mg; however, the volume of water in which it is dissolved is never of critical concern—water is even added to the Erlenmeyer flask while washing the wall during the titration. Explain why the addition of water to the Erlenmeyer flask is not critical to the analysis, whereas its addition to the NaOH solution is critical.

实验六　未知酸的测定

【实验目的】

(1) 掌握未知酸的测定原理和方法。

(2) 掌握酚酞指示剂滴定终点的判断方法。

【实验原理】

对于一未知酸溶液，可用NaOH标准溶液滴定，以酚酞指示剂指示滴定终点。由NaOH标准溶液的准确浓度和体积，可根据下式计算用于滴定的NaOH的物质的量：

$$n(\text{NaOH})=c(\text{NaOH}) \times V(\text{NaOH})$$

根据未知酸的类型（HA、H_2A或H_3A），从反应的化学计量数，可以计算反应中和的酸的物质的量n，结合准确量取的未知酸的体积V，计算未知酸的摩尔浓度c：

$$c=\frac{n}{V}$$

【仪器与试剂】

仪器：锥形瓶（250 mL×3）、移液管（25 mL）、滴定管。

试剂：未知酸溶液、酚酞指示剂、NaOH标准溶液。

【实验步骤】

(1) 准备3个干净的250 mL锥形瓶和1个干净的25 mL移液管。

(2) 移取25 mL未知酸溶液于250 mL锥形瓶中，加入2滴酚酞指示剂。

(3) 将滴定管清洗干净，用蒸馏水和NaOH标准溶液先后各润洗3次。往滴定管中加入NaOH标准溶液至近零刻度（注意排除管尖气泡），30 s后，读取并记录初始体积V_0。滴定未知酸溶液至酚酞变色并保持30 s，读取并记录滴定剂的最终体积V_1。

Experiment 6　Determination of an Unknown Acid Solution

【Objectives】

(1)　To master the principle and method of determining an unknown acid solution.

(2)　To master the color change of phenolphthalein at the endpoint.

【Principle】

In this experiment, an unknown molar concentration of an acid solution is determined. The standardized NaOH solution is used to titrate an accurately measured volume of the acid to the stoichiometric point. Phenolphthalein is employed to signal the endpoint of the titration. By knowing the molar concentration and volume of the NaOH Solution, the amount of substance of NaOH used for the analysis is:

$$n(NaOH)=c(NaOH) \times V(NaOH)$$

From the stoichiometry of the reaction (the instructor will inform students of the acid type, HA, H_2A, or H_3A), the amount of substance of acid neutralized in the reaction can be calculated. From the amount of substance of the acid that react and its measured volume, the molar concentration of the unknown acid is calculated:

$$c=\frac{n}{V}$$

【Apparatus and Chemicals】

Apparatus: Erlenmeyer flasks (250 mL × 3), volumetric pipet (25 mL), buret.

Chemicals: unknown acid solution, phenolphthalein indicator, standardized NaOH solution.

【Procedure】

(1)　Prepare three clean 250 mL Erlenmeyer flasks and a clean 25 mL volumetric pipet.

(2)　In an Erlenmeyer flask, pipet 25 mL of the acid solution. Add 2 drops of phenolphthalein.

(3)　Clean the base buret and rinse with three portions of the standardized NaOH solution. Fill the buret with the NaOH solution and after 30 s, read and record the initial volume. Titrate

（4）重复测定3份未知酸。计算未知酸的摩尔浓度和相对平均偏差。

【数据记录与处理】

将数据记录在表4.2中。

表4.2   未知酸的测定

项目	序号		
	I	II	III
$c(NaOH)/(mol \cdot L^{-1})$			
V_0/mL			
V_1/mL			
$V_1 - V_0$/mL			
$V(acid)$/mL	25.00		
$c(acid)/(mol \cdot L^{-1})$			
$\bar{c}(acid)/(mol \cdot L^{-1})$			
RAD/%			

【思考题】

在测定未知酸的摩尔浓度时，假设超过了化学计量点（滴定剂过量），这对酸的摩尔浓度有什么影响？为什么？

the acid sample to the phenolphthalein endpoint. After 30 s, read and record the final volume of titrant.

(4) Similarly titrate the other 2 aliquots of the acid solution. Calculate the average molar concentration of the unknown acid solution and the relative average deviation.

【Data Recording and Processing】

Record the experimental data in Table 4. 2.

Table 4. 2   Determination of an Unknown Acid Solution

Item	No.		
	I	II	III
$c(NaOH)/(mol \cdot L^{-1})$			
V_0/mL			
V_1/mL			
$V_1 - V_0/mL$			
$V(acid)/mL$	25.00		
$c(acid)/(mol \cdot L^{-1})$			
$\bar{c}(acid)/(mol \cdot L^{-1})$			
$RAD/\%$			

【Questions】

In the determination of the molar concentration of an unknown acid, suppose the stoichiometric point was surpassed (too much titrant was added to the analyte). What effect would this have on the reported molar concentration of the acid? Explain why.

实验七　食醋总酸度的测定

【实验目的】

(1) 熟练掌握滴定管、容量瓶和移液管的使用方法和滴定操作技术。

(2) 了解强碱滴定弱酸的反应原理及指示剂的选择。

(3) 学会食醋中总酸度的测定方法，了解食醋中醋酸的含量。

【实验原理】

食醋中的主要成分是醋酸（CH_3COOH，或简称HAc），此外还含有少量的其他有机弱酸如乳酸等，用NaOH标准溶液滴定食醋，在化学计量点时呈弱碱性，选用酚酞作指示剂。确切来说，在实验中我们测得的是食醋中的总酸量，而分析结果通常用含量最多的醋酸表示。

反应式：$CH_3COOH+NaOH \Longrightarrow CH_3COONa+H_2O$

计量关系：$n(CH_3COOH)=n(NaOH)$

【仪器与试剂】

仪器：锥形瓶（250 mL）、滴定管、移液管（25 mL）、容量瓶（250 mL）。

试剂：NaOH标准溶液（其配制和标定见实验五），0.2%酚酞指示剂，食醋。

【实验步骤】

准确移取25 mL食醋于250 mL容量瓶中，用新煮沸并冷却的蒸馏水稀释至刻度，摇匀。用移液管吸取上述试液25 mL于锥形瓶中，加25 mL蒸馏水稀释，加入2滴酚酞指示剂，摇匀，用已标定的NaOH标准溶液滴定至溶液呈浅粉红色，30 s内不褪色，即为终点。平行测定3份，根据NaOH标准溶液的浓度和滴定时消耗的体积，计算食醋的总酸度（单位：g/100 mL）。

Experiment 7   Determination of Total Acidity (Acetic Acid) in Vinegar

[Objectives]

(1) To skillfully master the use of burets, volumetric flasks and pipets as well as the titration technique.

(2) To comprehend the principle of the titration of weak acid with strong base and the selection of indicator.

(3) To learn the determination method of the total acidity in vinegar, and understand the content of acetic acid in vinegar.

[Principle]

Acetic acid (CH_3COOH) is the main active ingredient in vinegar and is responsible for its sour taste. Vinegars also contain a small amount of other weak organic acids, such as lactic acid. NaOH standard solution can be used for titration to measure acids in vinegar. At the stoichiometric point, the solution is weakly alkaline, and phenolphthalein is employed as the indicator. To be precise we should mention that we are determining not just acetic acid, but sum of all acids present in the vinegar. It is commonly expressed as acetic acid content.

The reaction of acetic acid with sodium hydroxide is:

$$CH_3COOH + NaOH == CH_3COONa + H_2O$$

Stoichiometric ratio: $n(CH_3COOH) = n(NaOH)$

[Apparatus and Chemicals]

Apparatus: Erlenmeyerflask (250mL), buret, pipet (25mL), volumetric flask (250mL).

Chemicals: NaOH standard solution (The preparation and standardization are described in Experiment 5), 0.2% phenolphthalein indicator, vinegar.

[Procedure]

Accurately pipet 25 mL of vinegar into a 250 mL volumetric flask. Dilute with newly boiled and cooled distilled water to the mark and shake well. Pipet 25 mL of the above solution into an Erlenmeyer flask, diluted with 25 mL of distilled water. Add 2 drops of phenolphthalein indicator, and shake well. Titrate with NaOH standard solution until the solution color changes from colorless to pink and does not fade within 30 seconds. Repeat the titration carefully two more times. According to the concentration of NaOH standard solution and the volume consumed in titration, calculate the total amount of acid in the vinegar (unit: g/100 mL).

【数据记录与处理】

将数据记录在表4.3中。

表4.3　食醋总酸度的测定

项目	序号		
	I	II	III
NaOH 始读数 V_0 / mL			
NaOH 终读数 V_1 / mL			
$V(\text{NaOH})=(V_1-V_0)$ / mL			
$c(\text{CH}_3\text{COOH})/(\text{mol}\cdot\text{L}^{-1})$			
$\bar{c}(\text{CH}_3\text{COOH})/(\text{mol}\cdot\text{L}^{-1})$			
$RAD/\%$			
$c(\text{CH}_3\text{COOH})/(\text{g/100 mL})$			

【注意事项】

（1）注意取食醋后应立即将试剂瓶盖好，防止挥发。

（2）食醋中醋酸浓度较高，且颜色较深，必须稀释后再测定。

（3）确保每次滴定前都重新装入NaOH溶液到近零刻度处。

【思考题】

（1）测定食醋总酸度时，为什么选用酚酞为指示剂？能否选用甲基橙或甲基红为指示剂？

（2）在滴定分析中，滴定管、移液管为什么需用操作溶液润洗3次？滴定中使用的锥形瓶或烧杯是否也要用操作溶液润洗？为什么？

（3）测定醋酸含量时，所用的蒸馏水不能含二氧化碳，为什么？

【Data Recording and Processing】

Record the experimental data in Table 4. 3.

Table 4. 3   Determination of Total Acidity（Acetic Acid）in Vinegar

Item	No.		
	I	II	III
NaOH initial reading V_0 / mL			
NaOH final reading V_1/ mL			
$V(NaOH)=(V_1-V_0)$/ mL			
$c(CH_3COOH)/(mol \cdot L^{-1})$			
$\bar{c}(CH_3COOH)/(mol \cdot L^{-1})$			
$RAD/\%$			
$c(CH_3COOH)/(g/100\ mL)$			

【Notes】

（1）Put the cap on the reagent bottle immediately after taking the vinegar to prevent volatilization.

（2）The acetic acid concentration in the vinegar is large, and the color is dark, so it must be diluted before titration.

（3）Make sure to refill the buret to near zero mark with the NaOH solution before each titration.

【Questions】

（1）Why is phenolphthalein used as the indicator when measuring the total acidity of vinegar? Can methyl orange or methyl red be used as indicators instead?

（2）In titration, why should burets and pipets be rinsed three times with solutions to be held or delivered? Should the Erlenmeyer flasks or beakers used in the titration be rinsed with solutions to be held? Why?

（3）When measuring the acetic acid content, the distilled water used cannot contain carbon dioxide. Why?

实验八　阿司匹林药片中乙酰水杨酸含量的测定

【实验目的】

（1）学习阿司匹林药片中乙酰水杨酸含量的测定。

（2）学习返滴定法的原理与操作。

【实验原理】

阿司匹林是最常用的药物之一，其主要成分乙酰水杨酸是有机弱酸（$K_a=1 \times 10^{-3}$），结构式 ![结构式 COOH OCOCH₃]，它的摩尔质量是180.16 g·mol^{-1}，微溶于水，易溶于乙醇。在强碱性溶液中可溶解并水解为水杨酸和乙酸盐，反应式如下：

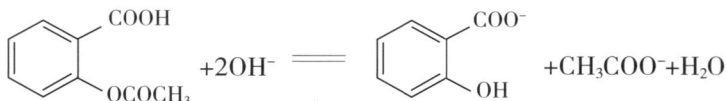

![反应式 COOH OCOCH₃ +2OH⁻ ⟶ COO⁻ OH + CH₃COO⁻ + H₂O]

由于药片中一般都添加一定量的赋形剂如硬脂酸镁、淀粉等不溶物，不宜直接滴定，可采用返滴定法进行测定。将药片研磨成粉末状后加入过量的NaOH标准溶液，加热一段时间使乙酰基水解完全，再用HCl标准溶液回滴过量的NaOH，滴定至溶液由红色变为接近无色即为终点。在这一滴定反应中，1 mol乙酰水杨酸消耗2 mol NaOH。

乙酰水杨酸若是纯品，可用NaOH溶液直接滴定，以酚酞为指示剂，滴定反应为：

![反应式 COOH OCOCH₃ +OH⁻ ⟶ COO⁻ OCOCH₃ + H₂O]

滴定应在10℃以下的中性乙醇介质中进行，以防止乙酰基水解。

【仪器与试剂】

仪器：滴定管、移液管、容量瓶、锥形瓶等。

试剂：NaOH溶液（500 mL，1 mol·L^{-1}）、HCl溶液（500 mL，0.1 mol·L^{-1}）、硼砂

Experiment 8   Determination of Acetylsalicylic Acid in Aspirin Tablets

【Objectives】

(1) To learn the determination of acetylsalicylic acid in aspirin tablets.

(2) To learn the principle and operation of back titration.

【Principle】

Aspirin was once widely used as an antipyretic and analgesic. Its main ingredient is acetylsalicylic acid. Acetylsalicylic acid is an organic weak acid ($K_a=1 \times 10^{-3}$) with structural

formula: , and its molar mass is $180.16 \text{g} \cdot \text{mol}^{-1}$. It is slightly soluble in water and readily soluble in ethanol. It can dissolve in a strongly alkaline solution and hydrolyze to salicylic acid and acetate as shown in the following reaction:

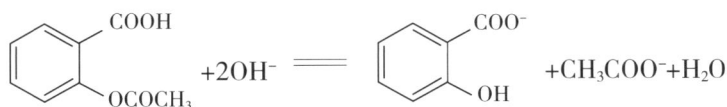

Since a certain amount of excipients such as magnesium stearate, starch and other insoluble substances are generally added in the tablets, it is not suitable to titrate aspirin tablets directly, so the back titration can be used for determination. The tablets are ground into a powder. A measured excess of NaOH standard solution is then added. It is heated up for a period of time to ensure complete hydrolysis of the acetyl group. Then, the excessive NaOH is back titrated with the HCl standard solution, and the endpoint is signaled by the color change from red to nearly colorless. In this titration reaction, 1 mole of acetylsalicylic acid consumes 2 moles of NaOH.

Acetylsalicylic acid can be titrated directly with the NaOH solution if it is pure. Phenolphthalein is employed as an indicator, and the titration reaction is:

The titration should be carried out in a neutral ethanol medium below 10℃ to prevent the hydrolysis of the acetyl group.

【Apparatus and Chemicals】

Apparatus: buret, pipet, volumetric flask, Erlenmeyer flask, etc.

（$Na_2B_4O_7 \cdot 10H_2O$）基准物、酚酞指示剂、甲基红指示剂、阿司匹林药片等。

【实验步骤】

1. $0.1 \ mol \cdot L^{-1}$ HCl溶液的标定

用减量法准确称取0.4~0.6 g硼砂，置于250 mL锥形瓶中，加50 mL水溶解，滴加2滴甲基红指示剂，用HCl溶液滴定至溶液由黄色变为浅红色即为终点。平行滴定3份。注意：在准备好进行滴定之前，不要溶解样品。计算HCl溶液的浓度及相对平均偏差。

2. 药片中乙酰水杨酸含量的测定

将阿司匹林药片研磨成粉末后，称量约0.6 g药粉，置入干燥小烧杯中，用移液管准确加入25 mL 1 $mol \cdot L^{-1}$ NaOH标准溶液，加入30 mL水，盖上表面皿，轻摇几下，置于近沸水浴加热15 min，迅速用流水冷却，将烧杯中的溶液定量转移到100 mL容量瓶中，用蒸馏水稀释至刻度，摇匀。

准确移取上述溶液10.00 mL于250 mL锥形瓶中，加蒸馏水20~30 mL，加入2~3滴酚酞指示剂，用$0.1 \ mol \cdot L^{-1}$ HCl标准溶液滴定至红色刚好消失即为终点。平行测定3份，根据所消耗的HCl溶液的体积计算药片中乙酰水杨酸的质量分数。

3. NaOH标准溶液与HCl标准溶液体积比的测定（空白实验）

用移液管准确移取25 mL 1 $mol \cdot L^{-1}$ NaOH标准溶液于小烧杯中，在与测定药粉相同的实验条件下进行加热，冷却后，定量转移至100 mL容量瓶中，稀释至刻度，摇匀。准确移取上述NaOH溶液10.00 mL于250 mL锥形瓶中，加水20~30 mL，加入2~3滴酚酞指示剂，用$0.1 \ mol \cdot L^{-1}$ HCl标准溶液滴定至终点，平行测定3份，计算$V(NaOH)/V(HCl)$的值。

【数据记录与处理】

将数据记录在表4.4至表4.6中。

表4.4　HCl标准溶液的标定

项目	序号		
	I	II	III
$m_{基准物}$/g			
HCl 初读数 V_0 /mL			
HCl 终读数 V_1/mL			
$V_1 - V_0$ /mL			
$c(HCl)/(mol \cdot L^{-1})$			
$\bar{c}(HCl)/(mol \cdot L^{-1})$			
RAD/%			

Chemicals: NaOH standard solution (500 mL, 1 mol·L^{-1}), HCl standard solution (500 mL, 0.1 mol·L^{-1}), borax ($Na_2B_4O_7·10H_2O$) primary standard, phenolphthalein indicator, methyl red indicator, aspirin tablets, etc.

[Procedure]

1. Standardization of 0.1 mol·L^{-1} HCl solution

Weigh accurately by different 0.4–0.6g of borax(in triplicate) in a 250 mL Erlenmeyer flask. Add 50 mL of water for dissolution. Add 2 drops of methyl red indicator. Titrate with the HCl solution until it changes from yellow to light red. Repeat the procedure with the other two primary standard samples. Note: Do not dissolve the samples until you are ready to perform the titration. Calculate the molar concentration of the HCl solution and the relative average deviation (RAD).

2. Determination of acetylsalicylic acid in tablets

After the grinding aspirin tablets into powder, weigh approximately 0.6 g of the powder into a dry small beaker. Add with pipet 25 mL of 1 mol·L^{-1} NaOH standard solution, followed by 30 mL of water. Cover the beaker with a watch glass, give it a few gentle shakes, and place it in a near-boiling water bath for 15 min. Quickly cool it with running water. Quantitatively transfer the solution from the beaker to a 100 mL volumetric flask. Dilute it with DI water to the mark and shake well.

Pipet 10.00 mL of the above solution into a 250 mL Erlenmeyer flask. Add 20–30 mL of DI water. Add 2–3 drops of phenolphthalein indicator. Titrate with HCl standard solution until the red color just disappears. Perform the determination in triplicate and calculate the mass fraction of acetylsalicylic acid in the tablets based on the volume of HCl solution consumed.

3. Determination of the volume ratio of NaOH standard solution to HCl standard solution (Blank experiment)

Pipet 25 mL of 1 mol·L^{-1} NaOH standard solution into a small beaker. Heat under the same experimental conditions as that in step 2. After cooling, quantitatively transfer it to a 100 mL volumetric flask. Dilute it to the mark and shake well. Pipet 10.00 mL of the above NaOH solution into a 250 mL conical flask. Add 20–30 mL DI water and then 2–3 drops of phenolphthalein indicator. Titrate with HCl standard solution to the endpoint and titrate three replicates. Calculate the value of $V(NaOH)/V(HCl)$.

[Data Recording and Processing]

Record the experimental data in Table 4.4, Table 4.5 and Table 4.6.

表4.5   药片中乙酰水杨酸含量的测定

项目	序号		
	I	II	III
m(乙酰水杨酸试样)/g			
V(移取试液)/ mL			
V(HCl)/ mL			
ω(乙酰水杨酸)/%			
$\bar{\omega}$(乙酰水杨酸)/%			
相对偏差 /%			
RAD/%			

表4.6   NaOH标准溶液与HCl标准溶液体积比的测定

项目	序号		
	I	II	III
V(NaOH)/ mL			
V(HCl)/ mL			
V(NaOH)/V(HCl)			
V(NaOH)/V(HCl)平均值			

【思考题】

(1) 在测定药粉的实验中，为什么1 mol乙酰水杨酸消耗2 mol NaOH，而不是3 mol NaOH？返滴定后的溶液中，水解产物的存在形式是什么？

(2) 列出计算药片中乙酰水杨酸含量的关系式。

(3) 若测定的是乙酰水杨酸的纯品（晶体），可否采用直接滴定法？

Table 4.4 Standardization of an HCl Solution

Item	No.		
	I	II	III
m_{prim_std} /g			
HCl initial reading V_0 /mL			
HCl final reading V_1/mL			
V_1-V_0 /mL			
$c(HCl)/(mol \cdot L^{-1})$			
$\bar{c}(HCl)/(mol \cdot L^{-1})$			
$RAD/\%$			

Table 4.5　Determination of Acetylsalicylic Acid in Tablets

Item	No.		
	I	II	III
$m(acetylsalicylic\ acid)/g$			
$V(pipeted\ solution)/mL$			
$V(HCl)/mL$			
$\omega(acetylsalicylic\ acid)/\%$			
$\bar{\omega}(acetylsalicylic\ acid)/\%$			
Relative deviation /%			
$RAD/\%$			

Table 4.6　Determination of the Volume Ratio Between NaOH Standard
Solution and HCl Standard Solution

Item	No.		
	I	II	III
$V(NaOH)/mL$			
$V(HCl)/mL$			
$V(NaOH)/V(HCl)$			
Average of $V(NaOH)/V(HCl)$			

【Questions】

(1) In the determination of acetylsalicylic acid in tablets, why does 1 mol of acetylsalicylic acid consume 2 mol of NaOH instead of 3 mol? What is the existing form of the hydrolytic product in the solution after back titration?

(2) Give a formula for calculating the content of acetylsalicylic acid in the tablets.

(3) Can direct titration be used if pure acetylsalicylic acid (crystal) is determined?

实验九　水溶液中 pH 值的测定（pH 计法）

【实验目的】

（1）理解pH计测定溶液pH值的原理。

（2）掌握pH计测定溶液pH值的方法。

【实验原理】

电位法是一种利用pH计测定溶液pH值的方法。一般是以玻璃电极为指示电极（−），饱和甘汞电极为参比电极（+），插入待测溶液中组成工作电池。电池电动势在给定温度下与溶液pH值呈线性关系。25 ℃时，溶液的pH值变化1个单位时，电池的电动势改变59.16 mV。

实际测量中，选用pH值与被测水样的pH值接近的标准缓冲溶液，校正（又叫定位）pH计，并保持溶液温度恒定，以减少由于液接电位、不对称电位及温度等变化而引起的误差。测定水样之前，用两种不同pH值的缓冲溶液校正，如用一种pH值的缓冲溶液定位好pH计后，再用该pH计测定相差约3个pH单位的另一种缓冲溶液的pH值，pH值误差应在±0.1之内。

校正后的pH计，可以直接测定水样或溶液的pH值。

【仪器与试剂】

仪器：pH计、pH玻璃复合电极、温度计、烧杯等。

试剂：邻苯二甲酸氢钾标准缓冲溶液（pH=4.00），磷酸二氢钾和磷酸氢二钠标准缓冲溶液（pH=6.86）、硼砂标准溶液（pH=9.18）、待测溶液、自来水、去离子水。

【实验步骤】

1. pH计的校正

参考第二章第三节酸度计部分。

Experiment 9   Determination of pH Value in Aqueous Solution（by pH Meter）

【Objectives】

(1) To comprehend the principle of measuring pH with a pH meter.

(2) To master the method of using a pH meter to measure the pH value of a solution.

【Principle】

The potentiometric method for determining solution pH involves immersing a glass electrode as the indicator electrode(−) and a saturated calomel electrode(SCE) as the reference electrode (+) into the test solution to form a working cell. The electromotive force (EMF) of this cell exhibits a linear relationship with the solution's pH at a given temperature. At 25 ℃, a change of 1 pH unit corresponds to an EMF shift of 59.16 mV.

In actual measurement, a standard buffer solution with a pH value close to that of the water sample is selected to calibrate the pH meter (also known as positioning), and the solution temperature is kept constant to reduce errors caused by liquid junction potential, asymmetric potential and temperature variations. Before measuring the water sample, two buffer solutions with different pH values are used for calibration. If the pH meter is positioned using one buffer solution and then used to measure the pH value of another buffer solution with a pH difference of approximately 3 units, the error should be within ± 0.1.

The calibrated pH meter can directly measure the pH value of the water sample or solution.

【Apparatus and Chemicals】

Apparatus：pH meter, pH glass combination electrode, thermometer, beaker, etc.

Chemicals：Standard buffer solution of $KHC_8H_4O_4$ （pH=4.00）, standard buffer solution of potassium dihydrogen phosphate and disodium hydrogen phosphate （pH=6.86）, borax standard solution （pH=9.18）, test solution, tap water, deionized water.

【Procedure】

1. Calibration of the pH meter

Refer to Section 3 of Chapter 2 on the pH meter.

2. Determination of the response slope of the pH glass electrode

Choose to measure E value. Immerse the electrode into the standard buffer solution with pH value of 4. 00. Swirl the beaker to ensure the solution is homogeneous. Read its E value on the display screen, and measure the E value of the standard solutions with pH values of 6.86 and

2. pH玻璃电极响应斜率的测定

选择测定电动势值E，将电极插入pH值为4.00的标准缓冲溶液中，摇动烧杯，使溶液均匀，在显示屏上读出其电动势值E，依次测量pH值为6.86和pH值为9.18的标准溶液的电动势值E。

3. 水样或溶液pH值的测定

用蒸馏水冲洗电极3～5次，再用被测水样或溶液冲洗电极3～5次，然后将电极放入水样或溶液中，待电极稳定，读出pH值。平行测定3次。

【数据记录与处理】

将数据记录在表4.7及表4.8中。

表4.7　pH玻璃电极响应斜率的测定

	标准缓冲溶液（pH=4.00）	标准缓冲溶液（pH=6.86）	标准缓冲溶液（pH=9.18）	自来水样	去离子水
E					
E–pH斜率					
pH					

表4.8　溶液pH值的测定

平行测定次数	Ⅰ	Ⅱ	Ⅲ	平均值
待测溶液的pH值				
水样的pH值				

【注意事项】

1. 玻璃电极的使用方法

（1）使用前，将玻璃电极的球泡部位浸在蒸馏水中24 h以上（传统电极）或2～10 h（现代电极）。复合电极应浸泡在与外参比液相同的溶液（如3 mol·L⁻¹ KCl或含KCl的pH=4.00的缓冲溶液）中。具体浸泡方法和时间，请参考电极说明书。

（2）以玻璃电极测定碱性水样或溶液时，应尽快测定。测量胶体溶液、蛋白质和染料溶液时，用后必须用棉花或软纸蘸乙醚小心地擦拭，再以酒精清洗，最后用蒸馏水洗净。

9.18 in sequence.

3. Determination of the pH value of water samples or solutions

Rinse the electrode with distilled water for 3 to 5 times, then rinse it with the water sample or solution to be measured for 3 to 5 times. Place the electrode in the water sample or solution, and read the pH value when the reading is stable. Perform three replicate tests.

[Data Recording and Processing]

Record the experimental result in Table 4. 7 and Table 4. 8.

Table 4. 7　Determination of the Response Slope of the pH Glass Electrode

	Standard buffer solution (pH=4.00)	Standard buffer solution (pH=6.86)	Standard buffer solution (pH=9.18)	Tap water	DI water
E					
E–pH Slope					
pH					

Table 4. 8　Determination of the pH Value of the Solution

Number of replicate measurement	I	II	III	Average
pH of the test solution				
pH of the water sample				

[Notes]

1. Use of glass electrode

(1) Before use, soak the bulb of the glass electrode in distilled water(24h for traditional models, 2–10 h for modern versions). Composite electrodes should be stored in solution matching their external reference electrode (e.g., 3 mol $\cdot L^{-1}$ KCl or KCl–containing pH 4.00 buffer). Refer to manufacturer's instructions for immersion protocols.

(2) When using the glass electrode to measure alkaline water samples or solutions, the measurement should be conducted promptly. After measuring colloidal solutions, proteins, and dye solutions, the electrode must be meticulously wiped with cotton or soft paper dipped in ether, cleaned with alcohol, and ultimately rinsed with distilled water.

2. Calibration of the pH meter

(1) A standard buffer solution that is proximate to the pH of the water sample should be selected to calibrate the pH meter.

(2) Preparation of pH standard buffer solutions is shown in Table 4.9. For pH primary standards purchased from reagent store, prepare them according to the instructions.

2. 仪器校正

（1）应选择与水样pH值接近的标准缓冲溶液校正仪器。

（2）pH标准缓冲溶液的配制见表4.9。试剂商店购买的pH基准试剂，按说明书配制。

表4.9   pH标准缓冲溶液的配制

序号	标准溶液浓度	pH（25 ℃）	1 000 mL蒸馏水中基准物质的质量/g
1	0.05 mol·L⁻¹ 二草酸三氢钾	1.679	12.61
2	饱和酒石酸氢钾（25 ℃）	3.559	6.4[①]
3	0.05 mol·L⁻¹ 柠檬酸二氢钾	3.776	11.41
4	0.05 mol·L⁻¹ 邻苯二甲酸氢钾	4.008	10.12
5	0.025 mol·L⁻¹ 磷酸二氢钾+ 0.025 mol·L⁻¹ 磷酸氢二钠	6.865	3.388[②]+3.533[②③]
6	0.008 695 mol·L⁻¹ 磷酸二氢钾+ 0.030 43 mol·L⁻¹ 磷酸氢二钠	7.413	1.179[②]+4.302[②③]
7	0.01 mol·L⁻¹ 四硼酸钠	9.180	3.80[③]
8	0.025 mol·L⁻¹ 碳酸氢钠+0.025 mol·L⁻¹ 碳酸钠	10.012	2.029+2.640
9	饱和氢氧化钙（25 ℃）	12.454	1.5[①]

注：①近似溶解度；②110 ℃～130 ℃烘干2 h；③用新煮沸并冷却的无CO_2蒸馏水。

【思考题】

（1）复合电极有哪些优缺点？使用前后应如何处理？为什么？

（2）用pH计测量pH值时，为什么必须用标准缓冲溶液校正仪器？

（3）pH计能否测定有色溶液或混浊溶液的pH值？

Table 4. 9　Preparation of pH Standard Buffer Solutions

No.	Concentration of standard solution	pH(25 ℃)	Mass of the primary standard in 1 000 mL distilled water/g
1	0.05 mol·L^{-1} potassium trihydrogen dioxalate	1.679	12.61
2	Saturated potassium hydrogen tartrate(25 ℃)	3.559	6.4[①]
3	0.05 mol·L^{-1} potassium dihydrogen citrate	3.776	11.41
4	0.05 mol·L^{-1} potassium hydrogen phthalate	4.008	10.12
5	0.025 mol·L^{-1} potassium dihydrogen phosphate + 0.025 mol·L^{-1} disodium hydrogen phosphate	6.865	3.388[②]+3.533[②③]
6	0.008 695 mol·L^{-1} potassium dihydrogen phosphate+ 0.030 43 mol·L^{-1} disodium hydrogen phosphate	7.413	1.179[②]+4.302[②③]
7	0.01 mol·L^{-1} sodium tetraborate	9.180	3.80[③]
8	0.025 mol·L^{-1} sodium bicarbonate+ 0.025 mol·L^{-1} sodium carbonate	10.012	2.029+2.640
9	Saturated calcium hydroxide(25 ℃)	12.454	1.5[①]

Note: ①Approximate solubility; ②Dried at 110 ℃ ~ 130 ℃ for 2 h; ③Using freshly boiled and cooled CO_2-free distilled water.

【Questions】

(1)　What are the advantages and disadvantages of a combination electrode? How should it be treated before and after use? And why?

(2)　Why must a standard buffer solution be used to calibrate the instrument when measuring pH with a pH meter?

(3)　Can a pH meter measure the pH value of a colored or turbid solution?

实验十　醋酸电离常数和电离度的测定

【实验目的】

(1) 学习测定弱酸的电离常数和电离度的方法。

(2) 进一步熟悉移液管、滴定管及容量瓶的使用方法。

(3) 学习pH计的使用。

【实验原理】

醋酸 （CH_3COOH或简写成HAc） 是弱电解质，在水溶液中存在如下电离平衡：

$$HAc \rightleftharpoons H^+ + Ac^-$$

$$K_a = \frac{[H^+][Ac^-]}{[HAc]} = \frac{[H^+]^2}{c-[H^+]}$$

$$\alpha = \frac{[H^+]}{c} \times 100\%$$

$[H^+]$、$[Ac^-]$和$[HAc]$分别为 H^+、Ac^-、HAc的平衡浓度，K_a为电离常数，α为电离度。

HAc溶液的总浓度c可以用标准NaOH溶液滴定测得，其电离出来的H^+离子的浓度可在一定温度下用pH计测定HAc溶液的pH值，根据pH=$-\lg[H^+]$计算出来，代入上述公式便可计算出该温度下的K_a值。

为使获得的实验结果较准确，可在一定温度下，测定一系列不同浓度的HAc溶液的pH值，然后将所得一系列相应的K_a值取平均值。

【仪器与试剂】

仪器：pH计、容量瓶 （50 mL）、吸量管 （10 mL）、碱式滴定管 （50 mL）、锥形瓶 （250 mL）、烧杯 （50 mL）。

试剂：NaOH标准溶液 （约0.1 mol·L^{-1}，标定方法见实验五）、HAc溶液 （0.1 mol·L^{-1}），酚酞指示剂 （2g·L^{-1}）。

Experiment 10   Determining the Dissociation Constant and the Degree of Dissociation of Acetic Acid

【Objectives】

(1) To learn the method of determining the dissociation constant and the degree of dissociation of weak acids.

(2) To further master the use of pipets, burets and volumetric flasks.

(3) To learn how to use a pH meter.

【Principle】

Acetic acid (CH$_3$COOH or HAc) is a weak electrolyte. It has the following dissociation equilibrium in aqueous solution:

$$HAc \rightleftharpoons H^+ + Ac^-$$

$$K_a = \frac{[H^+][Ac^-]}{[HAc]} = \frac{[H^+]^2}{c-[H^+]}$$

$$\alpha = \frac{[H^+]}{c} \times 100\%$$

$[H^+]$, $[Ac^-]$, and $[HAc]$ are the equilibrium concentrations of H^+, Ac^- and HAc respectively. K_a is the dissociation constant, and α is the degree of dissociation.

The analytical concentration of the acetic acid solution (c) can be determined by titration with standard NaOH solution. A pH meter is employed to measure the pH of the solution at a certain temperature and then calculate the concentration of the dissociated H^+ ions ($[H^+]$) according to pH= $-lg[H^+]$. The K_a value at this temperature can then be calculated by substituting c and $[H^+]$ into the above formula.

To obtain more accurate experimental results, the pH values of a series of acetic acid solutions with different concentrations can be determined at a certain temperature, and then the corresponding K_a values can be averaged.

【Apparatus and Chemicals】

Apparatus: pH meter, volumetric flask (50 mL), pipet (10 mL), base buret (50 mL), Erlenmeyer flask (250 mL), beaker (50 mL).

Chemicals: NaOH standard solution (about 0.1 mol·L^{-1}, see Experiment 5 for its standardization), HAc solution (0.1 mol·L^{-1}), phenolphthalein indicator (2 g·L^{-1}).

【实验步骤】

1. 用NaOH标准溶液测定HAc溶液的浓度(准确到三位有效数字)

用移液管吸取3份25 mL 0.1 mol·L^{-1} HAc溶液，分别置于锥形瓶中，各加2~3滴酚酞指示剂。分别用已标定的NaOH标准溶液滴定至溶液呈现微红色，30 s内不褪色为止。记录下所用NaOH溶液的体积。

2. 配制不同浓度的HAc溶液

用吸量管或滴定管分别取 5 mL、10 mL和25 mL已知其准确浓度的0.1 mol·L^{-1} HAc溶液于3个50 mL容量瓶中，用蒸馏水稀释至刻度，摇匀，制得0.01 mol·L^{-1}、0.02 mol·L^{-1}、0.05 mol·L^{-1} HAc溶液。

3. 测定0.01 mol·L^{-1}、0.02 mol·L^{-1}、0.05 mol·L^{-1}和0.1 mol·L^{-1} HAc溶液的pH值

用四个干燥的50 mL烧杯，分别取25 mL上述四种浓度的HAc溶液，由稀到浓分别用pH计测定它们的 pH值，并记录温度（室温）。

pH计的原理及其使用方法见第二章第三节。

【实验数据与处理】

自行设计实验记录和数据处理表格。

【思考题】

(1) 若改变所测溶液的温度，对结果是否有影响？

(2) 不同浓度的醋酸溶液的电离度是否相同？电离常数是否相同？

(3) 298 K时HAc的电离常数的文献值为1.754×10^{-5}，求本实验测定值的相对误差，并分析产生误差的原因。

(4) 除pH计法，还有哪些方法可以测定醋酸溶液的电离度和电离常数？简单说明其原理。

【Procedure】

1. Determining the concentration（to three significant digits）of acetic acid solution with the NaOH standard solution

Pipet three 25 mL aliquots of the 0.1 mol·L^{-1} HAc solution into three Erlenmeyer flasks, respectively. Add 2-3 drops of phenolphthalein indicator to each. Titrate each with the standardized NaOH solution until the solution is reddish and persists for half a minute. Record the volume of the NaOH solution used.

2. Preparing acetic acid solutions of different concentrations

Add with pipets or burets 5 mL, 10 mL and 25 mL of 0.1 mol·L^{-1} HAc solution with exactly known concentration into three 50 mL volumetric flasks, respectively. Dilute to the mark with distilled water and shake well to get 0.01 mol·L^{-1}, 0.02 mol·L^{-1} and 0.05 mol·L^{-1} HAc solutions.

3. Measuring the pH values of 0.01 mol·L^{-1}, 0.02 mol·L^{-1}, 0.05 mol·L^{-1} and 0.1 mol·L^{-1} HAc solutions

25 mL of the above four HAc solutions are added into four dry 50 mL beakers, and their pH values are measured from the dilute to the concentrated ones with a pH meter, and their temperature（room temperature）is recorded.

The principle and usage method of the pH meter are described in Section 3 of Chapter 2.

【Data Processing】

Design the experimental record and data processing tables by yourselves.

【Questions】

（1）If the temperature of the measured solution is changed, will it have an impact on the results?

（2）Are the dissociation degrees of acetic acid solutions of different concentrations the same? Are the dissociation constants the same?

（3）The literature value of the dissociation constant of HAc at 298K is 1.754×10^{-5}. Calculate the relative error of the measured value in this experiment and analyze the reasons for the error.

（4）In addition to applying the pH meter, what other methods can be used to determine the dissociation degree and constant of acetic acid solution? Briefly explain their principles.

实验十一　酸碱滴定设计实验

【实验目的】

（1）进一步掌握酸碱滴定的原理、有关计算方法及实验操作方法。

（2）通过认识酸碱滴定相关量关系、选择使用仪器和指示剂、讨论探索实验操作中有关问题，培养独立操作、分析问题和解决问题的能力。

（3）学习查阅参考文献及书写实验总结报告的方法。

【实验要求】

（1）学生应根据所选定的实验题目，查阅有关参考资料，并作详细记录。

（2）学生在查阅参考资料的基础上，拟订分析方案，经教师审阅后，进行实验工作，写出实验报告。

【设计内容】

分析方案的设计应包括方法原理、试剂配制、标准溶液的配制和标定、指示剂的选择、所需仪器的选择、取样量的确定、固体试样的溶样方法、具体的分析步骤以及分析结果的计算和讨论等。

【设计实验备选题】

1. 有机酸摩尔质量的测定

大多数有机酸是固体弱酸，如草酸（$pK_{a1} = 1.23$，$pK_{a2} = 4.19$）、酒石酸（$pK_{a1} = 2.85$，$pK_{a2} = 4.34$）、柠檬酸（$pK_{a1} = 3.15$，$pK_{a2} = 4.77$，$pK_{a3} = 6.39$）等，它们大多易溶于水，且有 $K_a \geqslant 10^{-7}$，即可在水溶液中用NaOH标准溶液进行滴定。反应产物为强碱弱酸盐，由于滴定突跃发生在弱碱性范围内，故常选用酚酞为指示剂，滴定至溶液呈微红色即为终点。根据NaOH标准溶液的浓度和滴定时所用的毫升数及称取的有机酸的质量，可计算该有机酸的摩尔质量。当有机酸为多元酸时，应根据每一级酸能否被准确滴定的

Experiment 11   Acid-Base Titration Design Experiment

【Objectives】

（1）To further master the principle, calculation and the experimental operation of acid-base titrations.

（2）To cultivate the ability to independently operate, analyze and solve problems through the stoichiometric relationship of acid-base reactions and titrations, the selection of apparatus and indicators, and the discussion and exploration of relevant problems in the experiment.

（3）To learn to consult references and write experimental summary reports.

【Requirements】

（1）Students should consult relevant reference materials according to the selected experiment topic and make detailed records.

（2）On the basis of the reference materials, students should draw up an analysis plan, and after being reviewed by the teacher, they can conduct experimental work and write experimental reports.

【Design Content】

The design of the analysis scheme should include the principle of the method, the preparation of reagents, the preparation and standardization of standard solutions, the selection of indicators, the selection of the required apparatus, the determination of the sampling amount, the dissolution method of solid samples, the specific analysis steps and the calculation and discussion of the analytical results.

【Recommended Topics】

1.   Determination of the molar mass of organic acids

Most organic acids are solid weak acids, such as oxalic acid $(pK_{a1} = 1.23, \quad pK_{a2} = 4.19)$, tartaric acid $(pK_{a1} = 2.85, \quad pK_{a2} = 4.34)$, citric acid $(pK_{a1} = 3.15, \quad pK_{a2} = 4.77, \quad pK_{a3} = 6.39)$, etc. Most of them are readily soluble in water, and have $K_a \geqslant 10^{-7}$, and thus can be titrated with NaOH standard solution in aqueous solution. The reaction product is a strong base-weak acid salt. Because the titration break occurs in the weakly alkaline range, phenolphthalein is often used as an indicator, and the endpoint color is slightly pink. The molar mass of the organic acid can be calculated according to the concentration and volume of the NaOH standard solution used in titration and the weight of the organic acid. When the organic acid is a polyacid, the quantitative

判别式（$c_{ai}K_{ai} \geq 10^{-8}$）及相邻两级酸之间能否分级滴定的判别式（$c_{ai}K_{ai}/c_{a(i+1)}K_{a(i+1)} \geq 10^5$）来判断多元酸与NaOH之间反应的计量关系，据此计算出有机酸的摩尔质量。

2. 氧化锌的测定

氧化锌是一种两性氧化物，不溶于水，难以直接滴定。但加入过量HCl溶解后，剩余的HCl可以甲基橙为指示剂，用NaOH标准溶液进行返滴定。

3. 饼干中Na_2CO_3、$NaHCO_3$含量的测定

含Na_2CO_3、$NaHCO_3$的混合碱可以HCl为标准溶液，分别用酚酞及甲基橙作指示剂来测定。当酚酞变色时，Na_2CO_3全部变为$NaHCO_3$，在此溶液中再加入甲基橙指示剂，继续滴定到终点即可。

4. $NaOH$–Na_3PO_4混合溶液中NaOH与Na_3PO_4浓度的测定

以百里酚酞为指示剂，用HCl标准溶液将NaOH滴定至NaCl，PO_4^{3-}滴定至HPO_4^{2-}。以甲基橙为指示剂，用HCl标准溶液将HPO_4^{2-}滴定至$H_2PO_4^-$。

5. 蛋壳中$CaCO_3$含量的测定

鸡蛋壳的主要成分为$CaCO_3$，还有$MgCO_3$、蛋白质、色素以及少量的Fe、Al。蛋壳中的碳酸盐能与过量的HCl标准溶液发生反应，过量的HCl溶液用NaOH标准溶液返滴定，由加入HCl的物质的量与返滴定所消耗的NaOH的物质的量之差，即可求得试样中$CaCO_3$的含量。

6. 乳酸钠注射液中乳酸钠含量的测定

乳酸钠注射液是注射用乳酸钠无菌水溶液，或由注射用无菌乳酸与NaOH反应制得，其水溶液碱性较弱，但是在乙酸溶液中碱性增强。可以结晶紫为指示剂，用高氯酸的乙酸标准溶液滴定测定其含量。此法属于非水滴定法，实验设计时要先配制和标定高氯酸的乙酸标准溶液。

7. 水泥熟料中二氧化硅含量的测定——氟硅酸钾法

氟硅酸钾滴定分析法属于间接酸碱滴定法。首先用碱熔融法将不溶于水和酸的试样中的SiO_2转化为可溶性硅酸盐，再在HNO_3溶液中与过量的K^+、F^-作用，定量生成难溶氟硅酸钾沉淀。将沉淀过滤，用KCl乙醇溶液洗涤后，用一定浓度的NaOH溶液中和未被洗净的残存酸，然后加入沸水使沉淀水解，生成与SiO_2相当量的HF，再用NaOH标准溶液滴定生成的HF。

relationship between the polyacid and NaOH should be determined according to the discriminant for whether each level of acid can be accurately titrated $(c_{ai}K_{ai} \geqslant 10^{-8})$ and the discriminant for dissociation step stepwise titration between adjacent dissociation steps $(c_{ai}K_{ai}/c_{a(i+1)}K_{a(i+1)} \geqslant 10^{5})$, and thereby the molar mass of the organic acid can be calculated.

2. Determination of zinc oxide

Zinc oxide is an amphoteric oxide, insoluble in water and difficult to titrate directly. However, after dissolution with an exactly known amount of excess HCl standard solution, the remaining amount of HCl can be back-titrated with a NaOH standard solution using methyl orange as an indicator.

3. Determination of NaHCO₃ and Na₂CO₃ content in cookies

The mixed base containing Na_2CO_3 and $NaHCO_3$ can be determined with HCl as the standard solution and phenolphthalein and methyl orange as the indicators respectively. When phenolphthalein changes color, all Na_2CO_3 has been transformed into $NaHCO_3$. In this solution, add methyl orange indicator and continue the titration until the endpoint is reached.

4. Determination of NaOH and Na₃PO₄ concentrations in NaOH–Na₃PO₄ mixed solutions

NaOH is titrated to NaCl, and PO_4^{3-} to HPO_4^{2-} with HCl standard solution using thymol-phthalein as indicator. After that methyl orange is added as an indicator, HPO_4^{2-} is titrated to $H_2PO_4^{-}$ with HCl standard solution.

5. Determination of CaCO₃ content in eggshell

The main component of eggshells is $CaCO_3$, followed by $MgCO_3$, proteins, pigments and trace amounts of Fe and Al. The carbonate in eggshells can react with a certain amount of excess HCl standard solution, and the excess HCl solution is back titrated with NaOH standard solution. The content of $CaCO_3$ in the sample can be obtained by the difference between the amount of HCl standard solution added and the amount of NaOH consumed in the back titration.

6. Determination of sodium lactate content in sodium lactate injection

Sodium lactate injection is a sterile solution of sodium lactate, or is prepared by the reaction of sterile lactic acid for injection with NaOH. Its aqueous solution is weakly alkaline. Its alkalinity is enhanced in acetic acid solution, and so it can be determined by titration with perchloric acid standard solution in acetic acid in the presence of crystal violet indicator. This method is a kind of the non-aqueous titration, and perchloric acid standard solution in acetic acid should be prepared and standardized before the experiment.

7. Determination of silica content in cement clinker—potassium fluosilicate method

Potassium fluosilicate titration method is an indirect acid-base titration. Initially, the SiO_2 in the sample, which is insoluble in water and acid, is transformed into soluble silicate through the alkali fusion method, and then precipitated quantitatively into insoluble potassium fluosilicate by

8. 食品添加剂中硼酸含量的测定

硼酸（H_3BO_3）的K_a=7.3×10^{-10}，故不能用NaOH标准溶液直接进行滴定。在H_3BO_3中加入甘油溶液使其生成甘油硼酸，其K_a=3×10^{-7}，可用NaOH标准溶液直接滴定。

interaction with excess K^+ and F^- in HNO_3 solution. After the precipitate is filtered and washed with KCl ethanol solution, the residual acid not removed by washing is neutralized with a specific concentration of NaOH solution. After that, the precipitate is hydrolyzed with boiling water to generate HF equivalent to SiO_2. Finally, the produced HF is titrated with NaOH standard solution.

8. Determination of boric acid content in food additives

The K_a of boric acid (H_3BO_3) is 7.3×10^{-10}, so it is too weak to be titrated directly with a NaOH standard solution. Glycerol solution is added to H_3BO_3 to produce glycerol boric acid $(K_a = 3 \times 10^{-7})$ which can be titrated directly with a NaOH standard solution.

第五章 配位滴定实验

Chapter 5 Complexometric Titration

实验十二　EDTA 溶液的标定

【实验目的】

(1) 掌握EDTA标准溶液的制备和标定方法。

(2) 掌握配位滴定原理，了解配位滴定的特点。

(3) 了解二甲酚橙指示剂的变色原理。

【实验原理】

乙二胺四乙酸（EDTA，H_4Y）是配位滴定中最常用的滴定剂，它能与大多数金属离子形成稳定的1:1配合物。EDTA微溶于水，故其标准溶液通常由乙二胺四乙酸二钠（$Na_2H_2Y \cdot 2H_2O$）来制备。EDTA试剂常吸附有少量水分并含有少量其他杂质，因此它的标准溶液通常采用间接法制备。常用于标定EDTA的基准物质有金属锌或镁、碳酸钙、金属铋等。为了减少测定误差，应尽量选择与被测金属离子相同的基准物，且使用与测定样品相同的方法来标定EDTA溶液。当用金属锌（Zn）来标定EDTA时，可选用二甲酚橙为指示剂，在pH=5~6的条件下滴定，终点颜色从紫红色变为黄色。

【仪器与试剂】

仪器：滴定管（50 mL）、锥形瓶（250 mL×3）、烧杯（150 mL）、容量瓶（250 mL）、移液管（25 mL）、量筒（10 mL）。

试剂：乙二胺四乙酸二钠（$Na_2H_2Y \cdot 2H_2O$，A. R）；金属锌（纯度99.99%）；HCl溶液（1:1）；六亚甲基四胺溶液（200 g/L）；二甲酚橙指示剂（2 g/L）。

【实验步骤】

1. 配制EDTA溶液

称取约1.9 g $Na_2H_2Y \cdot 2H_2O$于烧杯中，用蒸馏水溶解，倒入聚乙烯塑料瓶中，再加蒸馏水稀释至500 mL左右，摇匀。

Experiment 12   Standardization of EDTA Solution

【Objectives】

(1) To master the method of preparation and standardization of standard EDTA solution.

(2) To master the principle of complexometric titration and understand its characteristic.

(3) To learn to determine the endpoint of xylenol orange indicator.

【Principle】

Ethylenediamine tetraacetic acid （EDTA， H_4Y） is the most frequently utilized titrant in complexometric titration， forming stable 1 : 1 complexes with most metal ions. Due to its limited solubility in water， EDTA standard solutions are typically prepared by dissolving disodium dihydrogen ethylenediamine tetraacetate （$Na_2H_2Y \cdot 2H_2O$, also called EDTA） in water. Given that commercial EDTA reagents often contain impurities and adsorbed moisture， direct preparation of a precise concentration is challenging. Therefore， EDTA standard solutions are normally prepared using an indirect method and subsequently standardized against primary standards such as metallic zinc or magnesium， calcium carbonate， metallic bismuth and so on. For best results it is good to standardize EDTA solution against the same cation and using the same method as will be later used during sample analysis. When standardizing with metallic zinc （Zn）， xylenol orange serves as the indicator， and the titration is performed at pH 5–6. The endpoint is indicated by a color change from purplish-red to yellow.

【Apparatus and Chemicals】

Apparatus：buret （50 mL）， Erlenmeyer flasks （250 mL ×3）， beaker （150 mL）， volumetric flask （250 mL）， pipet （25.00 mL）， volumetric cylinder （10 mL）.

Chemicals：disodium dihydrogen ethylenediamine tetraacetate （$Na_2H_2Y \cdot 2H_2O$, A. R）； metallic zinc （purity 99.99%）; HCl solution （1 : 1）; methenamine （200 g/L）; xylenol orange solution （2 g/L）.

【Procedure】

1. Preparation of standard EDTA solution

Dissolve about 1.9 g of $Na_2H_2Y \cdot 2H_2O$ with distilled water to make 500 mL and mix well. Preserve it in a polyethylene bottle.

2. 配制250 mL 0.01 mol·L⁻¹ 锌标准溶液

称取0.155~0.172 g基准物锌片于干净的150 mL烧杯中，加入6 mL（1∶1）HCl溶液，立即盖上表面皿，待锌片完全溶解后，以少量蒸馏水冲洗表面皿，将溶液定量转移到250 mL容量瓶中，加蒸馏水至刻度，摇匀，计算Zn^{2+}标准溶液的浓度。

3. EDTA溶液（0.01 mol·L⁻¹）的标定

用移液管吸取 25.00 mL Zn^{2+}标准溶液于 250 mL 锥形瓶中，加2 滴二甲酚橙指示剂，滴加六亚甲基四胺溶液至溶液呈现稳定的紫红色，再加 5 mL 六亚甲基四胺溶液。然后用EDTA溶液滴定至溶液由紫红色刚好变为黄色，记下所用EDTA溶液的体积。平行滴定3份，计算EDTA溶液的准确浓度。

【数据记录与处理】

将实验数据记录在表5.1中。

表5.1 EDTA的标定

项目	序号		
	I	II	III
$m(Zn)$/g			
EDTA 初读数 V_0/mL			
EDTA 终读数 V_1/mL			
$V(EDTA)=(V_1-V_0)$/mL			
$n(Zn)$/mol①			
$c(EDTA)/(mol\cdot L^{-1})$②			
$\bar{c}(EDTA)/(mol\cdot L^{-1})$			
RAD/%			

注：①$n(Zn)=\dfrac{m(Zn)}{M(Zn)}\times\dfrac{1}{10}$；②$c(EDTA)=\dfrac{n(Zn)}{V(EDTA)}$。

【注意事项】

（1）EDTA在水中溶解缓慢，溶解时应搅拌或加热，以加速溶解；如有沉淀，应过滤溶液。

（2）EDTA溶液最好保存于聚乙烯塑料瓶中，也可以保存在硬质玻璃瓶中，特别是已经装过EDTA溶液的瓶子，以避免溶液与玻璃瓶中的金属离子发生反应。

（3）用稀酸溶液溶解锌时要小心操作，防止锌的流失。确保锌完全溶解，然后将溶

2. Preparation of 250 mL 0.01 mol·L⁻¹ zinc standard solution

Weigh accurately 0.155–0.172 g of the standard zinc into a 150 mL beaker. Add 6 mL of (1 : 1) HCl solution, and promptly cover it with a watch glass. After all the metal zinc is dissolved, rinse the watch glass and the inner wall of the beaker with a small quantity of distilled water. Quantitatively transfer the Zn^{2+} solution to a 250 mL volumetric flask. Dilute to the mark. Shake well for mixing. Calculate the molarity of the Zn^{2+} standard solution.

3. Standardization of EDTA solution (0.01 mol·L⁻¹)

Pipet 25.00 mL of the Zn^{2+} standard solution into a 250 mL Erlenmeyer flask. Add 2 drops of xylenol orange indicator. Add methenamine dropwise until the solution displays a stable purple-red color. Add again 5 mL of methenamine. Titrate with the EDTA solution until the color just changes from purple-red to yellow. Read and record the volume of the used EDTA solution. Repeat step 3 at least two more times and calculate the accurate concentration of the standard EDTA solution. Label on the bottle.

【Data Recording and Processing】

Record the experimental data in Table 5.1.

Table 5.1   Standardization of EDTA

Item	No.		
	I	II	III
$m(Zn)$/g			
EDTA initial reading V_0/mL			
EDTA final reading V_1/mL			
$V(EDTA)=(V_1-V_0)$/mL			
$n(Zn)$/mol①			
$c(EDTA)$/(mol·L⁻¹)②			
$\bar{c}(EDTA)$/(mol·L⁻¹)			
RAD/%			

Note: ①$n(Zn)=\dfrac{m(Zn)}{M(Zn)}\times\dfrac{1}{10}$; ②$c(EDTA)=\dfrac{n(Zn)}{V(EDTA)}$.

【Notes】

(1) EDTA dissolves slowly in water, so its solution should be shaken or warmed, in order to speed up the dissolution. If there is any residue, filtrate the solution.

(2) The EDTA solution may be kept in hard glass bottles; especially the bottles previously have kept EDTA solution in order to avoid the solution react with metal ions of the glass bottle. Polyethylene bottle is the most satisfactory for storage.

液定量转移到250 mL容量瓶中。

(4) 配位滴定反应缓慢。因此，在添加EDTA时，速度不能太快，尤其在较低的室温下要多加注意。接近终点时，逐滴加入滴定剂并剧烈摇动锥形瓶。

(5) 在配位滴定中，指示剂的体积是否合适是非常重要的，它会影响滴定终点。需要根据已有经验控制合适的指示剂用量。

(6) 掌握容量瓶、移液管和定量转移的操作。

【思考题】

(1) $Na_2H_2Y \cdot 2H_2O$有什么特点？

(2) 当以金属锌为基准物，以铬黑T为指示剂来标定EDTA溶液时，pH值应控制在哪个范围？可以使用六亚甲基四胺溶液作为缓冲溶液吗？在终点时颜色是如何变化的？

(3) 酸碱滴定和配位滴定有什么区别？我们应该注意哪些问题？

(4) 为什么EDTA溶液最好储存于聚乙烯塑料瓶中？

(5) 在用HCl溶解金属锌时，锥形瓶为什么要盖上表面皿？

(3)　Pay attention to operating when adding dilute acid solution to dissolve zinc so as not to lose any. Make sure zinc has dissolved completely before quantitatively transfering to a 250 mL volumetric flask.

(4)　The complexometric reaction proceeds slowly. Therefore, when adding EDTA, avoid rapid addition—especially at lower room temperatures—and exercise extra caution. Near the endpoint, add the titrant dropwise while vigorously swirling the Erlenmeyer flask.

(5)　In complexometric titration, the appropriate volume of the indicator is critical as it influences the endpoint. The optimal amount should be determined based on priorexperience.

(6)　Master the operations of volumetric flasks, pipets and quantitative transfer.

【Questions】

(1)　What are the properties of $Na_2H_2Y \cdot 2H_2O$?

(2)　When using zinc as the primary standard and eriochrome black T as the indicator to standardize the EDTA solution, what pH range should be maintained? Can methenamine be used as buffer solution? What is the color change at the endpoint?

(3)　What are the differences between acid-base titration and complexometric titration? Which problems should we pay attention to?

(4)　Why is it preferable to store the EDTA solution in polyethylene bottles?

(5)　When dissolving zinc with hydrochloric acid, why should a watch glass cover the Erlenmeyer flask?

实验十三　自来水硬度的测定

【实验目的】

(1) 掌握配位滴定法的原理。

(2) 掌握水的硬度的测定和计算方法。

(3) 学习水的硬度的意义和硬度的表达方法。

【实验原理】

水的硬度是衡量溶解在水中的钙和镁盐含量的概念。硬水不会对健康造成危害；然而，它会造成水垢，也会减少肥皂的起泡，降低洗涤剂的功效，它会破坏热水器和咖啡机等电器，也不能用于鱼缸。因此，简易快速测定水的硬度在实践中就显得特别重要。

配位滴定法是测定水的硬度的最佳方法之一。在pH值为10左右时，EDTA很容易与相同摩尔比（1∶1）的钙、镁反应。钙–EDTA配合物的稳定常数较镁–EDTA配合物的稍高，所以钙先反应，镁后反应。因此，滴定镁时所用的指示剂，即铬黑T，可以用来指示终点。如果水中不含镁，可以添加少量的镁–EDTA配合物（MgY^{2-}），再用铬黑T指示终点。加入三乙醇胺可以掩蔽其他金属离子如Fe^{3+}、Al^{3+}、Cu^{2+}、Pb^{2+}、Zn^{2+}的干扰。

溶液中如果含有碳酸盐，会干扰终点检测，可以用盐酸将溶液酸化，煮沸，然后用氨中和来去除。氨稍微过量不会造成影响，因为我们最后会加入氨缓冲溶液，而且pH值即使有零点几的微小变化也不成问题。

各国表示水硬度的方法不尽相同，表5.2为一些国家水硬度的换算关系。我国采用$c(CaCO_3)(mmol \cdot L^{-1})$或$c(CaCO_3)(mg \cdot L^{-1})$为单位表示水的硬度。

Experiment 13   Determination of Water Hardness

【Objectives】

(1) To master the principle of complexometric titration.

(2) To master the methods for determining and calculating water hardness.

(3) To understand the significance of water hardness and its expression methods.

【Principles】

Water hardness is a measure of the amount of calcium and magnesium salts dissolved in water. There is no health hazard associated with water hardness; however, hard water causes scale, as well as the reduced lathering of soaps. Hard water should not be used for washing (it reduces effectiveness of detergents) nor in water heaters and kitchen appliances like coffee makers (that can be destroyed by scale). It is also not good for fish tanks. In general, there are many applications where ability to easily determine water hardness is very important.

Complexometric titration is one of the best ways of measuring total water hardness. At pH around 10, EDTA easily reacts with both calcium and magnesium in the same molar ratio (1 : 1). The stability constant of the calcium-EDTA complex is a little bit higher than magnesium-EDTA complex, so calcium reacts first, followed by magnesium. Thus, for the endpoint, Eriochrome Black T, used in magnesium titration, can be employed. If the water contains no magnesium, a small amount of magnesium-EDTA complex (MgY^{2-}) is added to facilitate endpoint detection. Magnesium will be displaced by identical amount of calcium, and it will be titrated later, without affecting the final result. Triethanolamine can be added to mask the interference of other metal ions such as Fe^{3+}, Al^{3+}, Cu^{2+}, Pb^{2+} and Zn^{2+}.

If the solutions contain carbonates, they should be removed as they can interfere with endpoint detection. To do so, we can acidify the solution with HCl solution, boil it, and then neutralize with ammonia. Small excess of ammonia doesn't hurt, as we finally add ammonia buffer and change of pH by several tenths is not a problem.

Different countries employ different units to express water hardness. Table 5.2 shows the conversion relationship of water hardness units in some countries. In China, water hardness is represented by units of $c(CaCO_3)(mmol \cdot L^{-1})$ or $c(CaCO_3)(mg \cdot L^{-1})$.

表5.2　各国水硬度单位换算表

硬度单位	mmol·L^{-1}	德国硬度	法国硬度	英国硬度	美国硬度
1 mmol·L^{-1}	1.000 00	2.804 0	5.005 0	3.511 1	50.050
1德国硬度	0.356 63	1.000 0	1.784 8	1.252 1	17.848
1法国硬度	0.199 82	0.560 3	1.000 0	0.701 5	10.000
1英国硬度	0.284 83	0.798 7	1.425 5	1.000 0	14.255
1美国硬度	0.019 98	0.056 0	0.100 0	0.070 2	1.000

【仪器与试剂】

仪器：滴定管（50 mL）、锥形瓶（250 mL×3）、量筒（10 mL）、移液管（100 mL）。

试剂：EDTA标准溶液（0.01 mg·L^{-1}）、氨-氯化铵缓冲溶液（pH=10）、铬黑T（EBT）、三乙醇胺溶液、水样。

【实验步骤】

（1）制备水样：用移液管向250 mL锥形瓶中加入100.00 mL的水样，加入3 mL三乙醇胺溶液和5 mL氨-氯化铵缓冲溶液，再加入少量固体EBT指示剂。

（2）滴定：准备好滴定管，用EDTA标准溶液滴定水样，直到颜色从酒红色变为紫色再变为纯蓝色，记下所用EDTA溶液的初读数V_0和终读数V_1。接近终点时，反应缓慢，必须缓慢加入滴定液，并用力摇动锥形瓶。

（3）平行滴定3份，计算水样的总硬度，以$c(CaCO_3)(mg·L^{-1})$表示结果。计算平均值和相对平均偏差。

【数据记录与处理】

将实验数据记录在表5.3中。

Table 5.2   Conversion of Hardness Units in Various Countries

Hardness unit	$mmol \cdot L^{-1}$	German hardness	French hardness	British hardness	United States hardness
1 $mmol \cdot L^{-1}$	1.000 00	2.804 0	5.005 0	3.511 1	50.050
1 German hardness	0.356 63	1.000 0	1.784 8	1.252 1	17.848
1 French hardness	0.199 82	0.560 3	1.000 0	0.701 5	10.000
1 British hardness	0.284 83	0.798 7	1.425 5	1.000 0	14.255
1 United States hardness	0.019 98	0.056 0	0.100 0	0.070 2	1.000

【Apparatus and Chemicals】

Apparatus: buret (50 mL), Erlenmeyer flasks (250 mL×3), volumetric cylinder (10 mL), pipet (100 mL).

Chemicals: standard EDTA solution (0.01 $mg \cdot L^{-1}$), ammonia—ammonium chloride buffer solution (pH=10), eriochrome black T (EBT), triethanolamine solution, water sample.

【Procedure】

(1) Preparation of water sample: Add with a pipet a 100.00 mL aliquot of water to a 250 mL Erlenmeyer flask. Add 3 mL of triethanolamine solution and 5 mL of the ammonia-ammonium chloride buffer solution. Add a small amount of solid EBT indicator.

(2) Titration: Clean and fill the buret with the EDTA solution. Titrate until the color changes from wine red through purple to pure blue. Note the initial reading (V_0) and final reading (V_1) of the consumed EDTA solution. As the endpoint approaches, the reaction slows down. The titrant must be added slowly while vigorously swirling the Erlenmeyer flask.

(3) Refill the buret and repeat the titration at least two more times. Calculate and report the water hardness as $c(CaCO_3)(mg \cdot L^{-1})$. Calculate the average and the relative average deviation.

【Data Recording and Processing】

Record the experimental data in Table 5.3.

表5.3 水硬度的测定

项目	序号		
	Ⅰ	Ⅱ	Ⅲ
$c(EDTA)/(mol \cdot L^{-1})$			
V_0/mL			
V_1/mL			
$V(EDTA)=(V_1-V_0)/mL$			
$c(CaCO_3)/(mol \cdot L^{-1})$①			
水的硬度 $/(mg \cdot L^{-1})$②			
水的硬度平均值 $/(mg \cdot L^{-1})$			
$RAD/\%$			

注：①$c(CaCO_3)=\dfrac{c(EDTA) \times V(EDTA)}{V_{样本}}$；②水的硬度 $=c(CaCO_3) \times M(CaCO_3) \times 10^3$。

【注意事项】

（1）注意水样的采集时间、采集方法和采集设备。

（2）当滴定接近终点时，应剧烈摇动锥形瓶，放慢滴加的速度。

【思考题】

（1）水的硬度是什么意思？水的硬度单位是什么？

（2）在测定水的硬度时，结果应保留多少位有效数字？为什么？如果用量筒量取100 mL水样进行测定，结果应保留多少位有效数字？

（3）用EDTA溶液测定水的硬度时，水样中存在哪些离子会影响测定？如何避免它们的干扰？

（4）测定水硬度时，加入氨性缓冲溶液的目的是什么？在水硬度较高的样品中加入氨性缓冲溶液会出现什么异常现象？如何解决这个问题？

（5）如何分别测定水中Ca^{2+}和Mg^{2+}的含量？

（6）临时硬水和永久硬水的区别是什么？

Table 5.3 Determination of Water Hardness

Item	No.		
	I	II	III
$c(\text{EDTA})/(\text{mol} \cdot \text{L}^{-1})$			
V_0/mL			
V_1/mL			
$V(\text{EDTA}) = (V_1 - V_0)$/mL			
$c(\text{CaCO}_3)/(\text{mol} \cdot \text{L}^{-1})$ ①			
Water hardness②/$(\text{mg} \cdot \text{L}^{-1})$			
Average water hardness/$(\text{mg} \cdot \text{L}^{-1})$			
RAD/%			

Note: ① $c(\text{CaCO}_3) = \dfrac{c(\text{EDTA}) \times V(\text{EDTA})}{V_{\text{sample}}}$; ② Water Hardness $= c(\text{CaCO}_3) \times M(\text{CaCO}_3) \times 10^3$.

【Notes】

(1) Notice the collection time, method and apparatus for water samples.

(2) Near the endpoint, swirl the Erlenmeyer flask vigorously and slow down the titrant addition.

【Questions】

(1) What's the meaning of water hardness? Which ways can be used to express the units of water hardness?

(2) When determining the water hardness, how many significant figures should the result retain? Why? If a 100 mL water sample is measured with a graduated cylinder, how many significant figures should the result retain?

(3) Which ions exist in the solution can influence the determination of water hardness with EDTA solution? How to avoid it?

(4) What's the purpose of adding ammonia buffer when determining water hardness? What kind of abnormal phenomenon would appear after adding ammonia buffer in the sample with high water hardness? How to deal with it?

(5) How to determine the content of Ca^{2+} and Mg^{2+} in the water separately?

(6) What is the difference between temporary hard water and permanent hard water?

实验十四　铋、铅混合液中铋、铅含量的连续测定

【实验目的】

(1) 了解酸度对EDTA选择性滴定的影响。

(2) 掌握铋、铅连续测定的原理、方法和计算。

(3) 熟悉二甲酚橙指示剂的应用和终点颜色的变化。

【实验原理】

Bi^{3+}、Pb^{2+}均能与EDTA形成稳定的 $1:1$ 配合物，其稳定常数（$lg\,K$）分别为 27.94 和 18.04，二者稳定性差异显著（$\Delta lg\,K$=9.90>5）。因此，可通过控制溶液酸度的方法在同一份试液中连续滴定Bi^{3+} 和 Pb^{2+}，实现二者含量的分别测定。实验中均以二甲酚橙（XO）作指示剂，XO在 pH<6.3 时呈黄色，在 pH> 6.3 时呈红色；而它与Bi^{3+}、Pb^{2+}所形成的配合物都呈紫红色（分别记作Bi–XO、Pb–XO），且稳定性与Bi^{3+}、Pb^{2+}和EDTA所形成的配合物相比要低。此外，$K_{Bi-XO}>K_{Pb-XO}$。

测定时，先用 HNO_3 调节铅、铋混合液（下称混合液）至pH $\approx$ 1.0，此时，Bi^{3+} 与指示剂形成紫红色配合物（Pb^{2+}在此条件下不会与XO形成有色配合物），用EDTA标准溶液滴定至混合液由紫红色突变为亮黄色，即为滴定Bi^{3+}的终点。然后加入六亚甲基四胺溶液，使溶液 pH为 5~6，此时 Pb^{2+}与XO形成紫红色配合物，继续用EDTA标准溶液滴定至溶液由紫红色突变为亮黄色，即为滴定Pb^{2+}的终点。

【仪器与试剂】

仪器：滴定管（50 mL）、锥形瓶（250 mL×3）、量筒（10 mL）、移液管（25 mL）、烧杯。

试剂：EDTA标准溶液（0.01 mol·L⁻¹）；HNO_3溶液（0.10 mol·L⁻¹）；六亚甲基四胺溶

Experiment 14 Successive Titration of Bi^{3+} and Pb^{2+} in a Mixed Solution

【Objectives】

(1) To comprehend the effect of acidity on the selectivity of EDTA titration.

(2) To master the principle, method and calculation of successive determination of Bi^{3+} and Pb^{3+}.

(3) To be familiar with xylenol orange indicator and its color change at the endpoint.

【Principle】

Both Bi^{3+} and Pb^{2+} can form stable 1 : 1 complexes with EDTA, with the lg K values being 27.94 and 18.04 respectively. The stability disparity between them is considerable, with Δlg $K = 9.90 > 5$. Hence, it is feasible to conduct consecutive titrations of Bi^{3+} and Pb^{2+} in one test solution by regulating the acidity to determine their contents respectively. Xylenol orange (XO) is employed as the indicator. XO is yellow when pH < 6.3 and red when pH > 6.3. The indicator complexes of Bi^{3+} and Pb^{2+} (respeltively regarded as Bi–XO, Pb–XO) are purplish red with lower stability than those of the corresponding EDTA complexes. And $K_{Bi-XO} > K_{Pb-XO}$.

In the determination, the mixed solution is first adjusted to pH $\approx$ 1.0 with HNO_3. At this time, Bi^{3+} forms a purplish red complex with the indicator (Pb^{2+} cannot form a colored complex with XO under this condition). Titrate with the EDTA standard solution until it changes from purplish red to bright yellow, indicating the end of the titration of Bi^{3+}. Then add hexamethylenetetramine solution to make the pH 5–6, at which Pb^{2+} and XO form a purplish red complex. Continue to titrate with the EDTA standard solution until the solution changes from purplish red to bright yellow, which marks the endpoint of titrating Pb^{2+}.

【Apparatus and Chemicals】

Apparatus: Buret (50 mL), 3 of Erlenmeyer flasks (250 mL ×3), graduated cylinder (10 mL), pipet (25 mL), beaker.

Chemicals: 0.01 mol·L^{-1} EDTA standard solution; 0.10 mol·L^{-1} HNO_3; hexamethylene-tetramine solution (200 g·L^{-1}); Bi^{3+} and Pb^{2+} mixture (containing about 0.01 mol·L^{-1} Bi^{3+}, Pb^{2+} and 0.15 mol·L^{-1} HNO_3); XO (2 g·L^{-1}) aqueous solution.

液（200 g·L⁻¹）；Bi^{3+}、Pb^{2+}混合液（Bi^{3+}、Pb^{2+}含量各约为0.01 mol·L⁻¹，含0.15 mol·L⁻¹HNO_3）；XO（2 g·L⁻¹）水溶液。

【实验步骤】

（1）EDTA溶液的标定（同实验十二）。

（2）混合液中Bi^{3+}、Pb^{2+}含量的连续测定：用移液管移取25.00 mL混合液于250 mL锥形瓶中，加入10 mL 0.1 mol·L⁻¹ HNO_3、2滴XO，用EDTA标准溶液滴定至溶液由紫红色突变为亮黄色，即为滴定Bi^{3+}的终点，记下所用EDTA溶液的体积V_1（mL）。然后滴加200 g·L⁻¹六亚甲基四胺溶液至呈现稳定的紫红色，再多加入5 mL，此时溶液的pH为5~6。继续用EDTA标准溶液滴定至溶液由紫红色突变为亮黄色，即为滴定Pb^{2+}的终点，记下所消耗的EDTA的体积V_2（mL）。平行测定3份，计算混合液中Bi^{3+}和Pb^{2+}的含量（mol·L⁻¹）及V_1/V_2。

【数据记录与处理】

将实验数据记录在表5.4及表5.5中。

表5.4　混合液中铋的含量

项目	序号		
	Ⅰ	Ⅱ	Ⅲ
$V_{混合液}$/mL			
V_1/mL			
$c(Bi)/(mol·L^{-1})$			
$\bar{c}(Bi)/(mol·L^{-1})$			
RAD/%			

表5.5　混合液中铅的含量

项目	序号		
	Ⅰ	Ⅱ	Ⅲ
$V_{混合液}$/mL			
V_2/mL			
$c(Pb)/(mol·L^{-1})$			
$\bar{c}(Pb)/(mol·L^{-1})$			
RAD/%			

【Procedure】

(1) Standardization of EDTA solution (The same as Experiment 12).

(2) Successive determination of Bi^{3+} and Pb^{2+} in the mixed solution: Pipet 25.00 mL of the mixed solution into a 250 mL Erlenmeyer flask. Add 10 mL of $0.10 \ mol \cdot L^{-1}$ HNO_3 and 2 drops of XO. Then titrate with the EDTA standard solution until the color changes from purplish red to bright yellow, signaling the end of the titration of Bi^{3+}. Note the volume (V_1, mL). Then add $200g \cdot L^{-1}$ hexamethylenetetramine solution dropwise until the solution shows steady purplish red, and add an additional 5 mL. The pH of the solution is 5–6 now. Continue to titrate with EDTA standard solution until the solution changes from purplish red to bright yellow, which is the endpoint for titrating Pb^{2+}. Note the volume (V_2, mL). Titrate three replicate samples. Calculate the molar concentration ($mol \cdot L^{-1}$) of Bi^{3+} and Pb^{2+} in the mixture and V_1/V_2.

【Data Recording and Processing】

Record the experimental data in Table 5.4 and Table 5.5.

Table 5.4 Amount of Bismuth in the Mixture

Item	No.		
	I	II	III
$V_{mixture}$/mL			
V_1/mL			
$c(Bi)/(mol \cdot L^{-1})$			
$\bar{c}(Bi)/(mol \cdot L^{-1})$			
RAD/%			

Table 5.5 Amount of Lead in the Mixture

Item	No.		
	I	II	III
$V_{mixture}$/mL			
V_2/mL			
$c(Pb)/(mol \cdot L^{-1})$			
$\bar{c}(Pb)/(mol \cdot L^{-1})$			
RAD/%			

【注意事项】

(1) Bi^{3+}易水解，开始配制混合液时，所含HNO_3浓度较高，临使用前加水稀释至$0.15\ mol \cdot L^{-1}$左右。

(2) 测定Bi^{3+}时若酸度过低，Bi^{3+}将水解产生白色浑浊物，使终点过早出现，而且会产生回红现象。此时应放置片刻，继续滴定至溶液颜色变为稳定的透明亮黄色，即为终点。

(3) 滴定速度要慢，并且摇动锥形瓶以充分混匀。

【思考题】

(1) 描述连续滴定Bi^{3+}、Pb^{2+}过程中锥形瓶中颜色变化的情形，并说明颜色变化的原因。

(2) 能否取等量混合液两份，一份控制$pH \approx 1.0$滴定Bi^{3+}，另一份控制pH为$5\sim6$滴定Bi^{3+}、Pb^{2+}总量？为什么？

(3) 滴定Pb^{2+}时要调节溶液pH为$5\sim6$，为什么加入六亚甲基四胺溶液而不用NaOH、NaAc或$NH_3 \cdot H_2O$调节？

【Notes】

(1) To inhibit the hydrolysis of Bi^{3+}, the concentration of HNO_3 is high in the stock mixed solution, and it is diluted to about $0.15\ mol \cdot L^{-1}$ with water before use.

(2) When determining Bi^{3+}, if the acidity is too low, Bi^{3+} will hydrolyze to produce white turbidity, making the endpoint appear too early, and the solution return red. At this point, it should stand for a while and then be titrated with EDTA until the color of the solution becomes stable bright yellow, which is the endpoint.

(3) Titrate slowly and keep swirling the Erlenmeyer flask to ensure homogeneity.

【Questions】

(1) Please describe the color change in the Erlenmeyer flask during the successive titration of Bi^{3+} and Pb^{2+}, and explain why.

(2) Can two aliquots of the test solution be taken, where one portion has a pH of approximately 1.0 for titrating Bi^{3+}, and the other portion has a pH of 5−6 for titrating the total amount of Bi^{3+} and Pb^{2+}? Why or why not?

(3) When titrating Pb^{2+}, the pH of the solution needs to be adjusted to 5−6. Why is hexamethylenetetramine solution added instead of NaOH, NaAc or $NH_3 \cdot H_2O$?

实验十五　配位滴定设计实验

【实验目的】

(1) 培养运用配位滴定理论解决实际问题的能力，并通过实践加深对理论知识的理解。

(2) 培养查阅科技文献的能力。

(3) 培养自主设计实验方案及撰写实验报告的能力。

【实验要求】

(1) 在下面所罗列的题目中，选一个设计项目。

(2) 在查阅参考资料的基础上，拟订分析方案，经教师审阅后，进行实验工作，写出详细的实验报告。

(3) 实验方案大致有以下内容：测定方法概述；试剂的品种、数量和配制方法，试剂的浓度和体积；操作步骤；相关计算式和结果讨论等。

【设计实验备选题】

1. 硫酸铝中铝和硫的测定

用稀盐酸或稀硝酸溶解试样，用返滴定法测定铝。测定硫时，加过量Ba^{2+}后再用EDTA返滴定多余的Ba^{2+}。

2. 炉甘石中ZnO、$PbCO_3$、Fe_2O_3及（$CaCO_3+MgCO_3$）含量的测定

用酸溶解试样，控制酸度滴定Fe^{3+}、Zn^{2+}和Pb^{2+}、Ca^{2+}和Mg^{2+}、再用CN^-掩蔽（或解蔽）Zn^{2+}、Fe^{3+}、滴定Pb^{2+}。

3. 氨基酸钙片中钙含量的测定

氨基酸钙片中的钙以螯合型天门冬氨酸钙形式存在，由于EDTA–Ca稳定性比天门冬氨酸钙强，因此可在pH为10~11时采用EDTA标准溶液滴定螯合钙，以EBT+Mg–EDTA（或钙指示剂）为指示剂，实现对钙含量的测定。

Experiment 15   Complexometric Titration Design Experiment

【Objectives】

(1) To cultivate the ability to solve practical problems with the theory of complexometric titration, and deepen the understanding of theoretical knowledge through practice.

(2) To foster the ability to consult scientific and technological literature.

(3) To enhance the capacity to independently design experimental schemes and write experimental reports.

【Requirements】

(1) Choose one to design experiment project from the topics enumerated below.

(2) On the basis of consulting reference materials, draw up an analysis scheme. After it has been reviewed by the teacher, conduct the experimental work and write a detailed experimental report.

(3) The experimental scheme roughly contains the following contents: an overview of the determination method; varieties, quantities and preparation methods of reagents, as well as their concentrations and volumes; operation procedures; relevant calculation formulas and discussion of results.

【Recommended Topics】

1. Determination of aluminum and sulfur in aluminum sulfate

The sample is dissolved with dilute hydrochloric acid or dilute nitric acid solution, and then the aluminum is determined by back titration. When determining sulfur, an excess of standard Ba^{2+} is added to precipitate sulfate and the excess Ba^{2+} is back titrated with EDTA.

2. Determination of ZnO, $PbCO_3$, Fe_2O_3 and ($CaCO_3$+$MgCO_3$) content in calamine

The sample is dissolved in acid, titrated to determine Fe^{3+}, Zn^{2+} and Pb^{2+}, Ca^{2+} and Mg^{2+} by controlling the acidity. After masking Zn^{2+} and Fe^{3+} with CN^-, Pb^{2+} can be titrated.

3. Determination of calcium content in calcium amino acid tablets

The calcium in calcium amino acid tablets is in the form of chelated calcium aspartate. Since the stability of EDTA–Ca is stronger than that of calcium aspartate, the chelated calcium can be titrated with the EDTA standard solution at pH 10–11, and EBT+Mg–EDTA (or calcium indicator) is used as an indicator.

4. 食用盐中SO_4^{2-}含量的测定

在微酸性试液中，加入过量的$BaCl_2$，使硫酸根定量地与Ba^{2+}生成$BaSO_4$沉淀。过量的Ba^{2+}可在pH=10时，以EBT为指示剂，用EDTA溶液滴定，此时食盐中的Ca^{2+}、Mg^{2+}也同时被滴定，需要加以扣除。

5. Mg^{2+}–EDTA混合液中各组分的测定

在pH≈10的溶液中，以EBT为指示剂，检查哪种组分过量。

（1）若Mg^{2+}过量，移取一份试液用EDTA滴定过量的Mg^{2+}。另取一份试液调节pH至5~6，用XO作指示剂，用Zn^{2+}标准溶液滴定EDTA总量。

（2）若EDTA过量，移取一份试液调pH至5~6，用XO作指示剂，用Zn^{2+}标准溶液滴定EDTA总量。另取一份试液，加pH≈10的NH_3–NH_4Cl缓冲溶液，用EBT作指示剂，用Zn^{2+}标准溶液滴定过量的EDTA。

4. Determination of SO_4^{2-} in table salt

By adding excess barium chloride into the slightly acidic salt solution, the sulfate group is quantitatively precipitated as barium sulfate. The excessive barium can be titrated with the EDTA solution at pH=10 with EBT as the indicator. At this pH, Ca^{2+}、Mg^{2+} in the salt are also titrated and need to be deducted.

5. Determination of each component in Mg^{2+}–EDTA mixture

In a solution (pH ≈ 10), EBT is used as the indicator to check which component is excessive.

(1) If Mg^{2+} is in excess, pipet an aliquot of the test solution and titrate the excess Mg^{2+} with EDTA. Adjust the pH of another aliquot to 5–6, use XO as the indicator, and titrate the total amount of EDTA with the Zn^{2+} standard solution.

(2) If EDTA is in excess, an aliquot of the test solution is pipetted and adjusted to pH 5–6. With XO as the indicator, the total amount of EDTA is titrated with the Zn^{2+} standard solution. In another aliquot, NH_3–NH_4Cl buffer solution (pH ≈ 10) is added. With EBT as the indicator, the excessive EDTA is titrated with the Zn^{2+} standard solution.

第六章　氧化还原滴定实验

Chapter 6　Redox Titration

实验十六　硫代硫酸钠标准溶液的配制和标定

【实验目的】

（1）掌握硫代硫酸钠（$Na_2S_2O_3$）溶液的配制方法和保存条件。

（2）了解置换碘量法的过程、原理，并掌握用基准物$K_2Cr_2O_7$标定$Na_2S_2O_3$溶液浓度的方法。

（3）学习使用碘量瓶和掌握用淀粉指示剂指示滴定终点的方法。

【实验原理】

$Na_2S_2O_3 \cdot 5H_2O$一般都含有少量的杂质，如S、Na_2SO_3、Na_2SO_4、Na_2CO_3及NaCl等。它还容易风化和潮解，水中的CO_2、细菌和光照都能使其分解，水中的O_2也能将其氧化。因此，用于配制该溶液的水或储存溶液的玻璃器皿均需经过灭菌处理。若溶液出现浑浊、细菌或霉菌滋生等异常现象，整瓶溶液应予以废弃。基于上述特性，不能用直接法配制标准溶液，只能配制成近似浓度的溶液，然后标定其准确浓度。为减少溶解在水中的CO_2并杀死水中的微生物，应用新煮沸冷却后的蒸馏水配制溶液，并加入少量Na_2CO_3（使其在溶液中的浓度为0.02%）以防止$Na_2S_2O_3$的分解。$Na_2S_2O_3$溶液应贮存于棕色试剂瓶中，放置于暗处。7~14天后再进行标定，长期使用的溶液应定期标定。

标定$Na_2S_2O_3$溶液的基准物有$K_2Cr_2O_7$、KIO_3、$KBrO_3$和纯铜等，通常使用$K_2Cr_2O_7$基准物标定，标定时采用置换滴定法。$K_2Cr_2O_7$先与KI反应析出I_2：

$$Cr_2O_7^{2-} + 6I^- + 14H^+ \Longrightarrow 2Cr^{3+} + 3I_2 + 7H_2O$$

析出的I_2再用$Na_2S_2O_3$标准溶液滴定：

$$I_2 + 2S_2O_3^{2-} \Longrightarrow S_4O_6^{2-} + 2I^-$$

以淀粉溶液为指示剂，蓝色褪去即为终点。

Experiment 16   Standardization of Na$_2$S$_2$O$_3$ Solution

【Objectives】

(1) To master the preparation and preservation of Na$_2$S$_2$O$_3$ solution.

(2) To comprehend the process and principle of iodometry. Master the method to standardize Na$_2$S$_2$O$_3$ solution with the primary standard K$_2$Cr$_2$O$_7$.

(3) To learn to use iodine flasks and master the method of using starch indicator to signal the end of the titration.

【Principle】

Na$_2$S$_2$O$_3\cdot$5H$_2$O generally contains a small amount of impurities, such as S, Na$_2$SO$_3$, Na$_2$SO$_4$, Na$_2$CO$_3$ and NaCl. The decomposition of thiosulfate is generally attributed to the action of certain bacteria, light, and oxygen of the air catalyzed by traces of copper ions, and possibly carbon dioxide, which may change the molarity after a time. All water and glassware used to prepare and store the solution should be sterilized. If any turbidity or bacteria or mold growth appears, the solution should be discarded. Therefore, Na$_2$S$_2$O$_3\cdot$5H$_2$O standard solution can only be prepared to an approximate concentration, and then standardized to obtain its exact concentration. In order to reduce dissolved CO$_2$, O$_2$ and sterilize, the solution is prepared with freshly boiled and cooled distilled water, and a small amount of Na$_2$CO$_3$ is added (the concentration in the solution is 0.02%) to keep the solution neutral or slightly alkaline and thereby stabilize it against decomposition to elemental sulfur. The Na$_2$S$_2$O$_3$ solution should be stored in a brown reagent bottle in a dark place. The standardization will be carried out after 7–14 days. The solution used for an extended period should be standardized periodically.

Na$_2$S$_2$O$_3$ solution is standardized iodometrically against a pure oxidizing agent such as K$_2$Cr$_2$O$_7$, KIO$_3$, KBrO$_3$, or metallic copper. It is often standardized with K$_2$Cr$_2$O$_7$. In the standardization, iodine (triiodide) liberated by K$_2$Cr$_2$O$_7$ in an acidic potassium iodide solution is titrated with a Na$_2$S$_2$O$_3$. It is a displacement titration and the balanced equations for the standardization reactions are as follows:

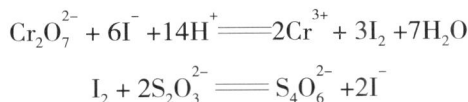

$$Cr_2O_7^{2-} + 6I^- + 14H^+ = 2Cr^{3+} + 3I_2 + 7H_2O$$
$$I_2 + 2S_2O_3^{2-} = S_4O_6^{2-} + 2I^-$$

【仪器与试剂】

仪器：电子天平、台秤、滴定管、移液管（25 mL）、容量瓶（250 mL）、碘量瓶（250 mL）、烧杯、棕色试剂瓶、量筒等。

试剂：$Na_2S_2O_3 \cdot 5H_2O$（固体）、Na_2CO_3（固体）、KI（200 g·L^{-1}）、$K_2Cr_2O_7$（基准物质）、HCl溶液（6 mol·L^{-1}），淀粉溶液（5 g·L^{-1}，配制方法：0.5 g淀粉，加少量水调成糊状，倒入100 mL沸腾的蒸馏水中，煮沸5 min冷却）。

【实验步骤】

（1）0.1 mol·L^{-1} $Na_2S_2O_3$溶液的配制：取500 mL蒸馏水，加热煮沸5~10 min，以确保无菌并排出二氧化碳，冷却至室温。称取12.5 g的$Na_2S_2O_3 \cdot 5H_2O$，放入500 mL棕色试剂瓶中，加入100 mL新煮沸冷却的蒸馏水，摇动使之溶解，等溶解完全后加入0.05 g Na_2CO_3，再加入400 mL新煮沸冷却的蒸馏水，摇匀，在暗处放置一周后，标定其浓度。

（2）$Na_2S_2O_3$溶液的标定：准确称取基准物$K_2Cr_2O_7$ 1.4 g于100 mL小烧杯中，加少量蒸馏水使之溶解，定量转移到250 mL容量瓶中，然后用蒸馏水稀释至刻度，摇匀。

准确移取25.00 mL $K_2Cr_2O_7$标准溶液于250 mL碘量瓶中，加5 mL 6 mol·L^{-1} HCl溶液和5 mL 200 g·L^{-1} KI，充分摇匀后盖好，放在暗处5 min。然后加100 mL蒸馏水，用0.1 mol·L^{-1} $Na_2S_2O_3$溶液滴定到呈浅黄绿色后（滴定过程应保持摇动锥形瓶，以免有过量的$Na_2S_2O_3$与酸性溶液接触），加入5 g·L^{-1}淀粉溶液2 mL，用少量蒸馏水冲洗锥形瓶内壁，继续滴定到蓝色消失而变为Cr^{3+}的绿色即为终点。重复步骤（2）再测定2份，为了减少实验误差，应每次在反应生成I_2后，马上用$Na_2S_2O_3$滴定。记录滴定前后，$Na_2S_2O_3$溶液的用量初读数V_0和终读数V_1。

（3）计算$Na_2S_2O_3$溶液的准确浓度和相对平均偏差。

【数据记录与处理】

将数据记录到表6.1中。

The endpoint is detected with starch. A sudden loss of the blue color indicates the end of the titration.

【Apparatus and Chemicals】

Apparatus: electronic balance, scale, buret, pipet (25 mL), volumetric flask (250 mL), iodine flask (250 mL), beaker, amber laboratory bottle, graduated cylinder.

Chemicals: $Na_2S_2O_3 \cdot 5H_2O$ (solid), Na_2CO_3 (solid), KI (200 g·L^{-1}), primary standard $K_2Cr_2O_7$, HCl solution (6 mol·L^{-1}), starch solution (5 g·L^{-1}, Preparation: Add a little water into 0.5 g starch to make paste. Pour into 100 mL of boiled distilled water and boil for 5 min, cool).

【Procedure】

(1) Preparation of 0.1 mol·L^{-1} $Na_2S_2O_3$ solution: Boil about 500 mL of distilled water for 5−10 min to ensure sterility and to expel carbon dioxide. Cool to room temperature. Clean a 500 mL amber laboratory bottle with cool boiled distilled water. Weigh 12.5 g of $Na_2S_2O_3 \cdot 5H_2O$ to the bottle, and then introduce 100 mL of freshly boiled and cooled distilled water and shake to dissolve. After complete dissolution, add 0.05 g of Na_2CO_3 and 400 mL of newly boiled and cooled distilled water, and shake thoroughly until the solution is homogeneous. After one week in the dark, standardize its concentration.

(2) Standardization of $Na_2S_2O_3$ solution: Weigh by difference 1.4 g of primary standard $K_2Cr_2O_7$ into a 100 mL beaker. Record the weight. Add distilled water into the beaker. Stir with a rod to dissolve the $K_2Cr_2O_7$. Quantitatively transfer it to a 250 mL volumetric flask. Invert and shake the volumetric flask for thorough mixing.

Pipet 25.00 mL of the standard $K_2Cr_2O_7$ solution into a 250 mL iodine flask. Add 5 mL of 6 mol·L^{-1} HCl solution. Add 5 mL of 200 g·L^{-1} KI solution. Swirl the Erlenmeyer flask. Stopper the flask and set the sample in the dark for about 5 min. Then add 100 mL of distilled water. Titrate with $Na_2S_2O_3$ solution until the solution becomes greenish-yellow, swirling the flask constantly so that no excess of $Na_2S_2O_3$ is in contact with the acidic solution at any time. Add 2 mL of 5 g·L^{-1} starch solution. Wash down the sides of the flask, and continue the titration until the blue color just disappears or a bright green color occurs. Individually repeat step (2) at least two more times. To minimize error, generate iodine (triiodide) just before being titrated with $Na_2S_2O_3$. Record the initial reading (V_0) and final reading (V_1) of the $Na_2S_2O_3$ solution volume before and after the titration.

(3) Calculate the molar concentration of the $Na_2S_2O_3$ solution. Calculate the relative average deviation.

【Data Recording and Processing】

Record the experimental data in Table 6.1.

表6.1 Na₂S₂O₃标准溶液的标定

项目	序号		
	I	II	III
$m(K_2Cr_2O_7)/g$			
V_0/mL			
V_1/mL			
$V(Na_2S_2O_3)=(V_1-V_0)$/mL			
$c(Na_2S_2O_3)/(mol \cdot L^{-1})$			
$\bar{c}(Na_2S_2O_3)/(mol \cdot L^{-1})$			
RAD/%			

【注意事项】

（1）$K_2Cr_2O_7$与KI的反应速度较慢，为了加快反应速度，可加6 mol·L⁻¹ HCl控制溶液酸度为0.2~0.4 mol·L⁻¹，同时加入过量KI后在暗处放置5 min（避光，防止KI被空气中的氧氧化），此反应才能定量完成。

（2）$Na_2S_2O_3$与I_2的反应只能在中性或弱酸性溶液中进行，因为在碱性溶液中会发生以下副反应：

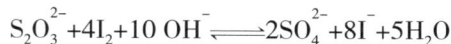

$$S_2O_3^{2-}+4I_2+10\,OH^- \Longrightarrow 2SO_4^{2-}+8I^-+5H_2O$$

而在酸性溶液中$Na_2S_2O_3$又易分解：

$$S_2O_3^{2-}+2H^+ \Longrightarrow S\downarrow+SO_2\uparrow+H_2O$$

所以滴定前溶液应加水稀释，一为降低酸度，二为使达到终点时溶液中的Cr^{3+}离子的绿色不会太深，影响观察终点。

（3）本实验以淀粉作指示剂，马铃薯淀粉比玉米淀粉更适于制作指示剂溶液，因其终点颜色变化更明显。淀粉溶液在有I^-离子存在时能与I_2分子形成蓝色可溶性吸附化合物，使溶液呈蓝色。达到终点时，溶液中的I_2全部与$Na_2S_2O_3$作用，则蓝色消失。淀粉指示剂并不是在滴定开始前加入，必须在滴定至近终点（稀释的碘色变成淡黄色）时方可加入，主要原因有两个：一是滴定开始时I_2太多，被淀粉吸附得过牢，就不易被完全夺出，难以观察终点；二是大多数碘量滴定法是在强酸介质中进行的，而淀粉在酸性溶液中有水解的倾向。

（4）若在终点有回褪现象，如果溶液不是很快变蓝，可认为是空气中氧的氧化作用造成，不影响结果；如果溶液变色很快且不断变蓝，说明溶液稀释过早，$K_2Cr_2O_7$与KI

Table 6.1　Standardization of $Na_2S_2O_3$ Solution

Item	No.		
	I	II	III
$m(K_2Cr_2O_7)/g$			
V_0/mL			
V_1/mL			
$V(Na_2S_2O_3)=(V_1-V_0)/mL$			
$c(Na_2S_2O_3)/(mol \cdot L^{-1})$			
$\bar{c}(Na_2S_2O_3)/(mol \cdot L^{-1})$			
$RAD/\%$			

【Notes】

(1)　$K_2Cr_2O_7$ reacts slowly with KI. To promote the reaction, 6 mol·L^{-1} HCl is added to control the solution at an acidity of 0.2–0.4 mol·L^{-1} and a large excess of iodide is added (common ion effect), and the solution is placed in a dark place for 5 min (to avoid light and prevent KI from being oxidized by oxygen in the air), so that the reaction can be completed quantitatively.

(2)　The reaction of $Na_2S_2O_3$ with I_2 can only take place in neutral or weakly acidic solutions.

In an alkaline solution the following side reaction occur:

$$S_2O_3^{2-}+4I_2+10\,OH^- \Longrightarrow 2SO_4^{2-}+8I^-+5H_2O$$

While in acidic solution, $Na_2S_2O_3$ is easy to decompose:

$$S_2O_3^{2-}+2H^+ \Longrightarrow S\downarrow +SO_2\uparrow +H_2O$$

Therefore, the solution should be diluted with water before titration, in order to reduce the acidity and to make the Cr^{3+} ions in the solution at the endpoint not too deep in green color, which will make it a little more difficult to determine the iodine-starch endpoint.

(3)　The endpoint for iodometric titrations is detected with starch. The disappearance of the blue starch-I_2 color indicates the end of the titration. Potato starch, rather than corn starch, is preferred for making the indicator solution since the color change due to the starch complex at the endpoint is sharper. The starch is not added at the beginning of the titration when the iodine concentration is high. Instead, it is added just before the endpoint when the dilute iodine color becomes pale yellow. There are two reasons for such timing. One is that the starch-I_2 complex is only slowly dissociated, and a diffuse endpoint would result if a large amount of the iodine were adsorbed on the starch. The second reason is that most iodometric titrations are performed in strongly acid medium and the starch has a tendency to hydrolyze in acid solution.

(4)　The endpoint color is not permanent and may fade back to blue in a matter of minutes, which should be disregarded. If the solution turns blue quickly and repeatedly, it indicates that the solution is diluted too early, and the $K_2Cr_2O_7$ and KI do not react completely. In this case,

作用不完全，需重新标定。

(5) 滴定开始时要慢摇快滴，但近终点时要慢滴，并用力振摇，防止吸附。

(6) 配制$Na_2S_2O_3$标准溶液所用的水，必须经过煮沸后放冷，以除去水中溶解的二氧化碳和氧，并杀灭微生物；在配制中还应加入0.02%的无水Na_2CO_3作为稳定剂，使溶液的pH值保持在9~10，以防止$Na_2S_2O_3$的分解。

(7) 为防止I_2的挥发，溶液不可受热。将碘瓶盖（或表面皿）取下后应用少许纯水将其表面附着物冲洗到瓶内。

【思考题】

(1) $Na_2S_2O_3$标准溶液如何配制？如何保存？

(2) 用$K_2Cr_2O_7$作基准物标定$Na_2S_2O_3$溶液浓度时，为什么要加入过量的KI和加入HCl溶液？为什么要将它放置在暗处一定时间？为什么在滴定前要加水稀释？如果①加KI不加HCl溶液；②加酸后不放置于暗处；③不放置或少放置一定时间即加水稀释，会产生什么影响？

(3) 滴定时，为什么必须在接近终点时加入淀粉指示剂？

(4) 写出用$K_2Cr_2O_7$溶液标定$Na_2S_2O_3$溶液的反应式和计算浓度的公式。

(5) 标定$Na_2S_2O_3$溶液的基准物有哪些？

the solution needs to be re-standardized.

(5) At the beginning of titration, it is necessary to swirl the Erlenmeyer flask slowly and drop the titrant fast. However, near the endpoint, the standard solution should be added dropwise and the Erlenmeyer flask should be swirled vigorously to prevent adsorption.

(6) In order to reduce dissolved CO_2, O_2 and sterilize, the solution is prepared with freshly boiled and cooled distilled water, and a small amount of Na_2CO_3 is added (the concentration in the solution is 0.02%) to keep the pH value of the solution at 9–10 and thereby stabilize it against decomposition to elemental sulfur.

(7) To avoid the volatilization of iodine, the solution should not be heated. After the reaction, the stopper (or watch glass) and the inner walls of the iodine flask should be rinsed with a little distilled water.

[Questions]

(1) How is the standard solution of $Na_2S_2O_3$ prepared and how is it preserved?

(2) In the standardization of $Na_2S_2O_3$ solution with $K_2Cr_2O_7$ as a primary standard, why should excess KI and HCl solution be added? Why should the solution be placed in dark for a certain period of time? Why is the solution diluted with water before titration?

What would be the implications if ①KI is added but no HCl solution; ②the solution is not placed in the dark after adding the acid; ③water is added for dilution without waiting or with only a short waiting period?

(3) Why isn't the starch indicator added at the beginning of the titration when the iodine concentration is high?

(4) Write the equations for the standardization of $Na_2S_2O_3$ solution with $K_2Cr_2O_7$ and the formula for calculating the concentration.

(5) What are the primary standards for the standardization of $Na_2S_2O_3$ solution?

实验十七　硫酸铜中铜含量的测定（碘量法）

【实验目的】

掌握用碘量法测定铜含量的原理和方法。

【实验原理】

在醋酸酸性介质（pH 3.5~4）中，Cu^{2+} 与过量KI作用生成CuI沉淀，同时析出定量的 I_2，反应如下：

$$2Cu^{2+} + 4I^- = 2CuI \downarrow + I_2$$

析出的 I_2 以淀粉为指示剂，可用 $Na_2S_2O_3$ 标准溶液滴定：

$$I_2 + 2S_2O_3^{2-} = 2I^- + S_4O_6^{2-}$$

Cu^{2+} 和 I^- 之间反应是可逆的，为使反应趋于完全，必须加入过量的KI；同时CuI沉淀强烈地吸附 I_2，易使测定结果偏低。通常采取的办法是在反应接近终点时加入KSCN溶液，将CuI沉淀吸附的 I_2 释放出来。

【仪器与试剂】

仪器：碘量瓶（250 mL）、容量瓶（250 mL）、滴定管（50 mL）、量筒（10 mL、100 mL）。

试剂：$Na_2S_2O_3$ 标准溶液（0.1 mol·L^{-1}）、HAc溶液（6 mol·L^{-1}）、KI溶液（20%）、淀粉溶液（0.5%）、$CuSO_4$·5H$_2$O 试样、KSCN溶液（10%）。

【实验步骤】

（1）称取 $CuSO_4$·5H$_2$O 试样5~7 g，用少量蒸馏水溶解后，定量转移到250 mL容量瓶，稀释至刻度，混匀。

（2）移取25 mL试样溶液于碘量瓶中，加入30 mL水，加4 mL 6 mol·L^{-1} HAc溶液，加10 mL KI溶液，立即用 $Na_2S_2O_3$ 标准溶液滴定至浅黄色，然后加入2 mL 0.5%淀粉溶液，

Experiment 17   Iodometric Determination of Copper

【Objectives】

To master the principle and method of determining copper by iodometry.

【Principle】

In acetic acid medium （pH 3.5–4）, Cu^{2+} reacts with excess KI to produce CuI precipitate, and an equivalence amount of I_2, as follows:

$$2Cu^{2+} + 4I^- =\!=\!= 2CuI \downarrow + I_2$$

The liberated I_2 is then titrated using a standard $Na_2S_2O_3$ solution with starch as an indicator and yields the subsequent reaction:

$$I_2 + 2S_2O_3^{2-} =\!=\!= 2I^- + S_4O_6^{2-}$$

The reaction between Cu^{2+} and I^- is reversible. Formation of the precipitate and the addition of excess iodide force the equilibrium to the right. It has been found that iodine is adsorbed onto the surface of the CuI precipitate and must be displaced to obtain correct results. KSCN solution is usually added just before the endpoint is reached to displace the adsorbed iodine.

【Apparatus and Chemicals】

Apparatus: iodine flask （250 mL）, volumetric flask （250 mL）, buret （50 mL）, graduated cylinder （10 mL, 100 mL）.

Chemicals: $Na_2S_2O_3$ standard solution （0.1 mol·L^{-1}）, HAc solution （6 mol·L^{-1}）, KI solution （20%）, starch solution （0.5%）, $CuSO_4·5H_2O$, KSCN solution （10%）.

【Procedure】

（1）weigh 5–7 g of the $CuSO_4·5H_2O$ sample. After dissolving it in a small amount of distilled water, transfer the solution quantitatively to a 250 mL volumetric flask. Make up the volume and shake well so as to affect uniform mixing.

（2）Pipet 25 mL of the sample solution into an iodine flask. Add 30 mL of water, 4 mL of 6 mol·L^{-1} HAc solution and 10 mL of KI solution. Immediately titrate to light yellow with the $Na_2S_2O_3$ standard solution. Then add 2 mL of 0.5% starch solution. Continue to titrate until a light blue color is achieved. Add 5 mL of KSCN solution. Swirl the flask gently and the solution turns to dark blue. Add the $Na_2S_2O_3$ standard solution dropwise until the blue color just disappears. The

继续滴定到呈浅蓝色，再加入5 mL KSCN溶液，摇匀后，溶液蓝色转深，继续用$Na_2S_2O_3$标准溶液滴定至蓝色刚好消失，此时溶液为米色CuSCN悬浮液，记录V（$Na_2S_2O_3$）的初始值V_0和终值V_1。

（3）平行测定三次（每次在滴定前才加入KI），根据滴定结果，计算$CuSO_4·5H_2O$试样中铜的百分含量。

【数据记录与处理】

将数据记录至表6.2。

$c(Na_2S_2O_3)=$ _____ $mol·L^{-1}$

$m_{铜盐}=$ _____ g

表6.2　硫酸铜中铜含量的测定

项目	序号		
	Ⅰ	Ⅱ	Ⅲ
V_0 / mL			
V_1 / mL			
$V(Na_2S_2O_3)=V_1-V_0$ / mL			
$c(Cu)/(mol·L^{-1})$			
$\omega(Cu)/\%$			
$\overline{\omega}(Cu)/\%$			
$RAD/\%$			

【注意事项】

（1）本实验中，I^-不仅是还原剂，也是Cu^+的沉淀剂，可使溶液中Cu^+大大降低，从而提高Cu^{2+}/Cu^+电对的电位，使氧化还原反应能定量进行。但Cu^{2+}与I^-反应是可逆的，为了使反应进行完全，必须加入过量的KI。

（2）为防止铜盐水解，反应必须在弱酸性溶液中进行。若酸度过低，则Cu^{2+}氧化I^-不完全，结果偏低而且反应速度慢、终点拖长。若酸度过高，则I^-被空气氧化为I_2，使Cu^{2+}的测定结果偏高。因此需用HAc或HAc–NaAc缓冲溶液控制溶液为弱酸性（pH 3.5~4）。

（3）在滴定过程中应不断摇动样品溶液。

（4）滴定时，溶液由棕红色变为土黄色，再变为淡黄色，表示已接近终点。

（5）KI应在滴定前再加入，且加入后，不必放置，应立即滴定，以防止CuI沉淀对I_2

solution is a beige CuSCN suspension. Record the consumed volume of $Na_2S_2O_3$ solution, take its initial reading(V_0)and final reading(V_1).

(3) Titrate two more replicates. Prepare and titrate the samples one at a time (note that potassium iodide should be added only before each titration). Report the result in terms of the percentage of copper (% Cu) in the unknown.

【Data Recording and Processing】

Record the experimental data in Table 6.2.

$c(Na_2S_2O_3)=$ _____ $mol \cdot L^{-1}$

$m(copper\ salt)=$ _____ g

Table 6.2 Determination of Copper Content in Copper Sulfate

Item	No.		
	I	II	III
V_0 / mL			
V_1/ mL			
$V(Na_2S_2O_3)= V_1-V_0$/ mL			
$c(Cu)/(mol \cdot L^{-1})$			
$\omega(Cu)/\%$			
$\bar{\omega}(Cu)/\%$			
$RAD/\%$			

【Notes】

(1) In this experiment, I^- is not only a reducing agent, but also a precipitant of Cu^+. It can reduce the concentration of Cu^+ greatly, thus increasing the potential of Cu^{2+}/Cu^+ redox pair, so that the redox reaction can be carried out quantitatively. However, the reaction between Cu^{2+} and I^- is reversible, and in order to complete the reaction, an excess of KI must be added.

(2) To prevent the hydrolysis of copper sulfate, the reaction must take place in a weakly acidic solution. If the acidity is too low, oxidation of I^- by Cu^{2+} is incomplete, the result is low, the reaction speed is slow and the endpoint is prolonged. If the acidity is too high, I^- is oxidized to I_2 by air, so that the determination result is high. Therefore, it is necessary to use HAc or HAc–NaAc buffer solution to control the solution to be weakly acidic (pH 3.5–4).

(3) The sample solution should be constantly swirled during titration.

(4) When titrating, the solution changes from brownish-red to earth-yellow and then to light yellow, indicating that the endpoint is approaching.

(5) KI should be added before titration. After KI is added, it should be titrated immediately to prevent the CuI precipitate from adsorbing I_2 too strongly.

的吸附太牢固。

(6) 加入KSCN可使CuI（$K_{sp}=1.1\times10^{-12}$）转化为溶解度更低的CuSCN沉淀（$K_{sp}=4.8\times10^{-15}$）。这样不但可以释放出被吸附的I_2，而且反应时再生出来的I^-与未反应的Cu^{2+}发生作用。在这种情况下，可以使用较少的KI而使反应进行得更完全。但KSCN只能在接近终点时加入，否则SCN^-可能直接还原Cu^{2+}而使结果偏低。

(7) 矿石或合金中的铜也可用碘量法测定。但必须设法防止其他能氧化I^-的物质（如NO_3^-、Fe^{3+}等）的干扰。防止的方法包括加入掩蔽剂以掩蔽干扰离子（例如使Fe^{3+}生成FeF_6^{3-}而被掩蔽）或在测定前将干扰离子分离除去。若有As（V）、Sb（V）存在，应将pH调至4，以免它们氧化I^-。

【思考题】

(1) 用碘量法测定铜含量时，加入KSCN的目的是什么？

(2) 为什么测定反应一定要在弱酸性溶液中进行？

(3) 碘量法有两个重要的误差来源，一是I_2的挥发，二是I^-被空气中O_2氧化。实验中应采取哪些措施减少这两种误差？

(4) 给出铜盐中铜的质量百分比的计算公式。

(6) If KSCN is added, CuI ($K_{sp}=1.1 \times 10^{-12}$) is converted to CuSCN precipitate ($K_{sp}=4.8 \times 10^{-15}$) with less solubility. In this way, not only can the adsorbed I_2 be liberated, but also the regenerated I^- will interact with the unreacted Cu^{2+}. And so less KI can be used to allow the reaction to proceed more completely. However, KSCN can only be added near the endpoint, otherwise SCN^- may directly reduce Cu^{2+}, resulting in a lower result.

(7) Copper in ores or alloys may also be measured iodometrically. However, some measures must be taken to prevent the interference of other substances that can oxidize I^- (such as NO_3^-, Fe^{3+}, etc.). This can be prevented by adding masking agents to mask the interfering ions (For example, Fe^{3+} is masked by making FeF_6^{3-}) or by separating and removing them before determination. If As(V) and Sb(V) are present, the pH should be adjusted to 4 to prevent them from oxidizing I^-.

[Questions]

(1) What is the purpose of adding KSCN in the iodometric determination of copper?

(2) Why must the determination reaction be carried out in a weakly acidic solution?

(3) The iodometric method has two important sources of error, one is the volatilization of I_2 and the other is the oxidation of I^- by O_2 in air. What measures should be taken to reduce these two errors in the experiment?

(4) Give the formula for calculating the percentage of copper by mass in copper salt.

实验十八   间接碘量法测定漂白剂中的次氯酸钠

【实验目的】

掌握漂白剂中次氯酸钠含量的测定和计算方法。

【实验原理】

漂白剂的主要成分是次氯酸钠（NaClO），NaClO在醋酸酸性溶液中与过量KI反应，释放出一定量的碘，再以$Na_2S_2O_3$标准溶液滴定，以淀粉指示剂确定终点。反应式如下：

$$ClO^- + 3I^- + 2H^+ == Cl^- + I_3^- + H_2O$$

$$I_3^- + 2S_2O_3^{2-} == S_4O_6^{2-} + 3I^-$$

【仪器与试剂】

仪器：滴定管（50 mL）、锥形瓶（250 mL×3）、量筒（10 mL）、移液管（25 mL）。

试剂：$Na_2S_2O_3$溶液（0.1 mol·L^{-1}）、KI（200 g·L^{-1}）、冰醋酸、淀粉溶液、漂白剂溶液（称取260.1 g市售漂白剂，并将之稀释至3L）。

【实验步骤】

移取25.00 mL漂白剂溶液到250 mL锥形瓶中，用蒸馏水稀释至100 mL。加入10 mL冰醋酸、5 mL 20% KI。立即用$Na_2S_2O_3$标准溶液滴定至淡黄色，加入2 mL淀粉溶液，继续滴定至无色（蓝色刚好褪去），记录$Na_2S_2O_3$标准溶液滴定前后的初值V_0和终值V_1。平行测定三份，计算漂白剂中NaClO的重量百分比和相对平均偏差。

【数据记录与处理】

将实验数据记录至表6.3中。

Experiment 18　Iodometric Determination of NaClO in Bleach

【Objectives】

To master the method of determination and calculation of NaClO content in bleach.

【Principle】

The main ingredient in bleach is NaClO. The oxidizing power (percent NaClO) of the solution is determined iodometrically by reacting it with an excess of iodide in acetic acid solution and titrating the iodine produced (I_3^- in the presence of excess iodide) with standard sodium thiosulfate solution, and a starch indicator is used.

Equations:

$$ClO^- + 3I^- + 2H^+ = Cl^- + I_3^- + H_2O$$

$$I_3^- + 2S_2O_3^{2-} = S_4O_6^{2-} + 3I^-$$

【Apparatus and Chemicals】

Apparatus: buret (50 mL), Erlenmeyer flasks (250 mL×3), graduated cylinder (10 mL), pipet. (25 mL)

Chemicals: $Na_2S_2O_3$ solution ($0.1\ mol \cdot L^{-1}$), KI ($200\ g \cdot L^{-1}$), glacial acetic acid, starch solution, bleach solution (weigh 260.1 g of commercially available bleach and dilute it to 3 L).

【Procedure】

Pipet 25.00 mL of the bleach solution into a 250 mL Erlenmeyer flask. Dilute to 100 mL with distilled water. Add 10 mL of glacial acetic acid. Add 5 mL of 20% KI. Titrate immediately with the standard $Na_2S_2O_3$ solution, swirling the flask constantly. When the color has faded to a pale yellow, add about 2 mL of starch solution and continue the titration until the solution just becomes colorless (the blue color just disappears). Record the consumed volume of $Na_2S_2O_3$ solution, take the initial reading(V_0)and final reading(V_1). Perform three parallel determination. Calculate the percentage by weight of NaClO in the solution and the relative average deviation (RAD).

【Data Recording and Processing】

Record the experimental data in Table 6.3.

表6.3 漂白剂中NaClO的测定

项目	序号		
	I	II	III
V_0 / mL			
V_1 / mL			
$V(Na_2S_2O_3) = V_1 - V_0$ / mL			
$c(Na_2S_2O_3)/(mol \cdot L^{-1})$			
$V(NaClO)$/L	25.00 mL = 25.00 × 10^{-3} L		
$M(NaClO)/(g \cdot mol^{-1})$	74.442		
NaClO/%			
NaClO 平均值 /%			
RAD/%			

【注意事项】

（1）滴定必须在弱酸性溶液中进行。

（2）加入的KI必须过量，一般KI过量2~3倍。

（3）滴定时不要剧烈摇动溶液，滴定速度宜适当快些，减少I_2的挥发和I^-被O_2氧化。

【思考题】

（1）在漂白剂溶液中加入冰醋酸和20% KI有什么作用？加入以上两种试剂后，溶液的颜色是什么？为什么？

（2）滴定终点的颜色变化是什么？为什么？

（3）如何计算漂白剂中NaClO的重量百分比？

Table 6. 3   Determination of NaClO in Bleach

Item	No.		
	I	II	III
V_0 / mL			
V_1 / mL			
$V(Na_2S_2O_3)= V_1-V_0$/ mL			
$c(Na_2S_2O_3)/(mol \cdot L^{-1})$			
$V(NaClO)$/L	25.00 mL = 25.00 × 10^{-3} L		
$M(NaClO)/(g \cdot mol^{-1})$	74.442		
NaClO/%			
NaClO 平均值 /%			
RAD/%			

【Notes】

(1) Titration must be performed in weakly acidic solutions.

(2) The added KI must be excessive, usually 2–3 times in excess.

(3) Swirl the Erlenmeyer flask gently during the titration; the titration speed should be appropriately fast to reduce the volatilization of I_2 and oxidation of I^- by O_2.

【Questions】

(1) What's the function of glacial acetic acid and 20% KI added to the bleach solution? What's the color of the solution after the addition of the above two reagents? Why?

(2) What's the color change at the titration endpoint? Why?

(3) How to calculate the percentage by weight of NaClO in the bleach?

实验十九　高锰酸钾标准溶液的配制和标定

【实验目的】

（1）了解$KMnO_4$标准溶液的配制方法和保存条件。

（2）掌握以$Na_2C_2O_4$作基准物标定$KMnO_4$标准溶液的原理和方法。

（3）了解$KMnO_4$作为自身指示剂的特点。

【实验原理】

市售的$KMnO_4$试剂常含有少量MnO_2和其他杂质，如硫酸盐、氯化物及硝酸盐等；另外，蒸馏水中常含有少量的有机物质，能使$KMnO_4$还原，且还原产物能促进$KMnO_4$自身分解，见光分解更快。因此，$KMnO_4$的浓度容易改变，不能用直接法配制准确浓度的$KMnO_4$标准溶液。$KMnO_4$溶液必须得到正确的配制和保存，如果长期使用，必须定期进行标定。

可用于标定$KMnO_4$的基准物质较多，有As_2O_3、$H_2C_2O_4 \cdot 2H_2O$、$Na_2C_2O_4$和纯铁丝等。其中以$Na_2C_2O_4$较易获得高纯度（99.95%）的基准品而最为常用，$Na_2C_2O_4$不含结晶水，不易吸湿，易制纯，性质稳定。用$Na_2C_2O_4$标定$KMnO_4$的反应为：

$$2MnO_4^- + 5C_2O_4^{2-} + 16H^+ = 2Mn^{2+} + 10CO_2\uparrow + 8H_2O$$

滴定在强酸性条件下进行，常用H_2SO_4调节酸度。$KMnO_4$也可作为自身指示剂，滴定时可利用MnO_4^-本身的紫红色指示反应终点。

【仪器与试剂】

仪器：台秤、分析天平、表面皿、小烧杯、大烧杯（1 000 mL）、电炉、棕色细口瓶、微孔玻璃漏斗、称量瓶、锥形瓶（250 mL×3）、量筒、聚四氟乙烯滴定管。

试剂：$KMnO_4$（A. R.）、$Na_2C_2O_4$基准物（于105 ℃干燥2 h后备用）、H_2SO_4（3 mol·L^{-1}）。

Experiment 19   Preparation and Standardization of KMnO$_4$ Solution

【Objectives】

(1) To comprehend the preparation and preservation of KMnO$_4$ standard solution.

(2) To master the principle and method of using Na$_2$C$_2$O$_4$ primary standard to standardize KMnO$_4$ solution.

(3) To understand the signaling of the endpoint by KMnO$_4$ itself.

【Principle】

KMnO$_4$ reagents often contain small amounts of MnO$_2$ and other impurities, such as sulfates, chlorides and nitrates. In addition, distilled water often contains a small amount of organic substances, which can reduce KMnO$_4$, and the reduction products can promote the decomposition of KMnO$_4$ itself. Light catalyzes the decomposition reaction. Therefore, the concentration of KMnO$_4$ is easy to change, and the KMnO$_4$ standard solution cannot be prepared by the direct method. Great caution must be taken to prepare and store the KMnO$_4$ solution. If it is used for a long time, it must be standardized regularly.

Many primary standards can be used to standardize KMnO$_4$, such as As$_2$O$_3$, H$_2$C$_2$O$_4\cdot$2 H$_2$O, Na$_2$C$_2$O$_4$ and pure iron wire. Na$_2$C$_2$O$_4$ is preferred most as available in a higher standard of purity (99.95%). It's available in the anhydrous form. Once dried it is practically nonhygroscopic. The reaction of standardizing KMnO$_4$ with Na$_2$C$_2$O$_4$ is:

$$2MnO_4^- + 5\ C_2O_4^{2-} + 16H^+ =\!=\!= 2Mn^{2+} + 10CO_2\uparrow + 8\ H_2O$$

The titration is carried out in strongly acidic medium, and H$_2$SO$_4$ is often used to control the acidity. KMnO$_4$ also acts as self-indicator as its slight excess gives a distinct pink color to the solution.

【Apparatus and Chemicals】

Apparatus: scale, analytical balance, watch glass, small beaker, large beaker (1 000 mL), hot plate, dark-colored bottle, microporous glass funnel, weighing bottle, Erlenmeyer flasks (250 mL$\times$3), graduated cylinder, PTFE buret.

Chemicals: KMnO$_4$ (A.R.), Na$_2$C$_2$O$_4$ primary standard (dried at 105 ℃ for 2 h before use), H$_2$SO$_4$ (3 mol$\cdot$L^{-1}).

【实验步骤】

1. KMnO₄标准溶液的配制

称量1.6 g固体KMnO₄，置于大烧杯中，加水至500 mL（由于要煮沸使水蒸发，可适当多加些水），盖上表面皿，煮沸约1 h，冷却后将溶液转移至棕色瓶内，在暗处放置2~3天，然后用微孔玻璃漏斗或玻璃棉漏斗过滤，滤液装入棕色细口瓶中，贴上标签，一周后标定。

2. KMnO₄标准溶液的标定

称取0.15~0.20 g基准物$Na_2C_2O_4$三份，分别置于250 mL的锥形瓶中，加约60 mL水和15 mL 3 mol·L^{-1} H_2SO_4，盖上表面皿，水浴加热到75 ℃~85 ℃（刚开始冒蒸气的温度），趁热用KMnO₄溶液滴定。开始滴定时反应速度慢，待溶液中产生了Mn^{2+}后，滴定速度可适当加快，但临近终点时滴定速度要减慢，同时不断摇动锥形瓶，直到溶液呈现微红色并维持30 s不褪色即终点。根据$Na_2C_2O_4$的质量和消耗KMnO₄溶液的体积计算KMnO₄浓度。用同样方法滴定其他两份$Na_2C_2O_4$溶液，相对平均偏差应在±0.2%以内。

【数据记录与处理】

将数据记录到表6.4中。

表6.4 KMnO₄标准溶液的标定

项目	序号		
	Ⅰ	Ⅱ	Ⅲ
$m(Na_2C_2O_4)$/g			
V_0/ mL			
V_1/ mL			
$V(KMnO_4)=(V_1-V_0)$/ mL			
$c(KMnO_4)$/(mol·L^{-1})			
$\bar{c}(KMnO_4)$/(mol·L^{-1})			
RAD/%			

【注意事项】

（1）蒸馏水中常含有少量还原性物质，可使KMnO₄还原为$MnO_2·nH_2O$。市售KMnO₄内含的细粉状的$MnO_2·nH_2O$能加速KMnO₄的分解，故通常将KMnO₄溶液煮沸一段时间，

【Procedure】

1. Preparation of KMnO$_4$ solution

Weigh 1. 6 g of solid KMnO$_4$ and place it in a large beaker. Add water to 500 mL （add more water as appropriate because the water must be boiled to evaporate）. Cover a watch glass and boil in a water-bath for about 1 h. After cooling, transfer the solution to a brown bottle, store in a dark place for 2–3 days, and then filter through a microporous glass or glass wool funnel. The filtrate is put into a brown glass bottle，labeled，and standardized after a week.

2. Standardization of KMnO$_4$ solution

Three replicates of 0.15–0.20 g of Na$_2$C$_2$O$_4$ primary standard are weighed and placed in 250 mL Erlenmeyer flasks. Add about 60 mL of water and 15 mL of 3 mol·L^{-1} H$_2$SO$_4$. Cover a watch glass，heat in a water-bath to 75 ℃–85 ℃ （the temperature at which steam just starts to rise）, titrate against KMnO$_4$ solution from the buret while hot. The reaction rate is slow at the beginning of titration，and the titration speed can be accelerated appropriately after Mn^{2+} is produced in the solution. But in the vicinity of the endpoint，the titration speed should be slowed down while constant swirling the contents until a faint pink color persists for 30 s. The concentration of KMnO$_4$ is calculated according to the mass of Na$_2$C$_2$O$_4$ and the volume of the consumed KMnO$_4$ solution （Take initial reading V_0 and final reading V_1）. Titrate the other two portions of the Na$_2$C$_2$O$_4$ solution in the same way，and the relative average deviation should be within 0.2%.

【Data Recording and Processing】

Record the experimental data in Table 6.4.

Table 6.4   Standardization of KMnO$_4$Standard Solution

Item	No.		
	I	II	III
$m(Na_2C_2O_4)/g$			
V_0/ mL			
V_1/ mL			
$V(KMnO_4)=(V_1-V_0)/$ mL			
$c(KMnO_4)/(\ mol·L^{-1})$			
$\bar{c}(KMnO_4)/(\ mol·L^{-1})$			
RAD/%			

【Notes】

（1） Distilled water often contains a small amount of reducing substances that can reduce KMnO$_4$ to MnO$_2$·nH$_2$O. The fine powder MnO$_2$·nH$_2$O contained in the commercial KMnO$_4$ can accelerate the decomposition of KMnO$_4$，so the KMnO$_4$ solution is usually boiled for a period of

冷却后，还需放置2~3天，使之充分作用，然后将沉淀物过滤除去。

（2）在室温条件下，$KMnO_4$与$C_2O_4^{2-}$之间的反应缓慢，故加热提高反应速度。但温度又不能太高，如温度超过85℃则有部分$H_2C_2O_4$分解，反应式如下：

$$H_2C_2O_4 \Longrightarrow CO_2\uparrow + CO\uparrow + H_2O$$

（3）$Na_2C_2O_4$溶液的酸度在开始滴定时，约为$1\,mol\cdot L^{-1}$，到达滴定终点时，约为$0.5\,mol\cdot L^{-1}$，这样能促使反应正常进行，并且防止MnO_2的形成。滴定过程如果产生棕色浑浊（MnO_2），应立即加入H_2SO_4补救，使棕色浑浊消失。

（4）刚开始滴定时，反应很慢，在第一滴$KMnO_4$还没有完全褪色以前，不可加入第二滴。当反应生成能使反应加速进行的Mn^{2+}后，可以适当加快滴定速度，但过快则可能导致局部$KMnO_4$过浓而分解，放出O_2或引起杂质的氧化，都可造成误差。

如果滴定速度过快，部分$KMnO_4$将来不及与$Na_2C_2O_4$反应，而会按下式分解：

$$4MnO_4^- + 4H^+ \Longrightarrow 4MnO_2 + 3O_2\uparrow + 2\,H_2O$$

（5）$KMnO_4$标准溶液滴定时的终点较不稳定，当溶液出现微红色，在30 s内不褪色时，就可认为滴定已经完成，如对终点有疑问，可先将滴定管读数记下，再加入1滴$KMnO_4$标准溶液，出现紫红色即证实已到达终点。

（6）滴定终点时溶液的温度不能低于60 ℃。

（7）由于$KMnO_4$溶液颜色很深，液面凹下弧线不易看出，因此应该从液面最高边上读数。

【思考题】

（1）在配制$KMnO_4$标准溶液时应注意哪些问题？

（2）在用$Na_2C_2O_4$标定$KMnO_4$标准溶液的过程中，加入酸、加热和控制滴定速度的目的是什么？

（3）为什么用H_2SO_4控制溶液的酸度？可以用HNO_3或HCl调节酸度吗？

（4）$KMnO_4$溶液的配制过程中要用微孔玻璃漏斗过滤，能否换用定量滤纸过滤，为什么？

（5）烧杯或锥形瓶等容器放置$KMnO_4$溶液较久后，其壁上会产生棕色沉淀物，这种沉淀是什么？如何去除此沉淀？

time. After cooling, it needs to be left for 2 to 3 days to allow full reaction, and then the sediment is filtered out.

(2) The reaction between $KMnO_4$ and $C_2O_4{}^{2-}$ is slow at room temperature, therefore, heating is required to increase the reaction rate. However, the temperature should not be too high. Above 85 ℃, $H_2C_2O_4$ begins to decompose according to the following reaction:

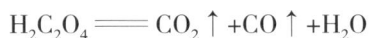

$$H_2C_2O_4 \Longrightarrow CO_2 \uparrow + CO \uparrow + H_2O$$

(3) The acidity of the $Na_2C_2O_4$ solution is about $1\ mol \cdot L^{-1}$ at the beginning of the titration and $0.5\ mol \cdot L^{-1}$ at the end of the titration. This allows the reaction to proceed normally and prevents MnO_2 from forming. If brown turbidity (MnO_2) is produced during the titration, H_2SO_4 should be added immediately to remove it.

(4) At the beginning of titration, the reaction is slow, and the $KMnO_4$ should be added dropwise with particular care to allow each drop to be fully decolorized before the next is introduced. When the reaction produces Mn^{2+}, which speeds up the reaction, the titration speed can be accelerated appropriately. However, if the titration is too fast, local $KMnO_4$ may be too concentrated and decomposed, which releases O_2 and oxidizes impurities and causes errors.

If the titration speed is too fast, some $KMnO_4$ will decompose according to the following equation instead of reacting with $Na_2C_2O_4$:

$$4MnO_4^- + 4H^+ \Longrightarrow 4MnO_2 + 3O_2 \uparrow + 2H_2O$$

(5) The endpoint in the titration of $KMnO_4$ standard solution is unstable. When the solution appears reddish and does not fade within 30 s, the titration can be considered to have been completed. If in doubt about the endpoint, the buret reading can be recorded first, and then another 1 drop of $KMnO_4$ standard solution can be added, the occurrence of purplish red confirms the end of titration.

(6) The temperature of the solution at the end of the titration must not be lower than 60 ℃.

(7) For highly colored solution such as $KMnO_4$, read the volume at the top edge of the meniscus, as the concave curve is difficult to discern.

【Questions】

(1) What should be paid attention to when preparing the $KMnO_4$ standard solution?

(2) What are the purposes of adding acid, heating and controlling titration speed during the standardization of the $KMnO_4$ solution with $Na_2C_2O_4$?

(3) Why is H_2SO_4 used to control the acidity? Can HNO_3 or HCl be used instead?

(4) During the preparation of the $KMnO_4$ solution, a microporous glass funnel should be used for filtration. Can quantitative filter paper be used instead? Why or why not?

(5) What is the brown precipitate on the walls of the containers such as beakers or conical flasks containing the $KMnO_4$ solution after being placed for a long time? How can this precipitate be removed?

实验二十　补钙制剂中钙含量的测定

【实验目的】

（1）了解沉淀分离的基本要求及操作。

（2）掌握KMnO₄法间接测定钙含量的原理及方法。

【实验原理】

利用某些金属离子（如碱土金属离子、Pb^{2+}、Cd^{2+}等）与草酸根能形成难溶的草酸盐沉淀的反应，可以用KMnO₄法间接测定这些金属离子的含量。如测定Ca^{2+}的反应如下：

$$Ca^{2+}+C_2O_4^{2-} =\!=\!= CaC_2O_4\downarrow$$

$$CaC_2O_4+2H^+ =\!=\!= Ca^{2+}+H_2C_2O_4$$

$$5H_2C_2O_4+2MnO_4^-+6H^+ =\!=\!= 2Mn^{2+}+10CO_2\uparrow+8H_2O$$

用该法可测定某些补钙制剂（如葡萄糖酸钙、钙立得、盖天力等）中的钙含量，分析结果是否与标示量吻合。

【仪器与试剂】

仪器：分析天平、电热板、烧杯、水浴锅、漏斗、量杯、酸式滴定管、锥形瓶、洗瓶。

试剂：KMnO₄溶液（$0.02\ mol\cdot L^{-1}$，配制及标定同实验十九）、$(NH_4)_2C_2O_4$溶液（$0.05\ mol\cdot L^{-1}$）、氨（NH_3）水（$7\ mol\cdot L^{-1}$）、HCl溶液（$6\ mol\cdot L^{-1}$）、H_2SO_4溶液（$1\ mol\cdot L^{-1}$）、甲基橙水溶液（$1\ g\cdot L^{-1}$）、$AgNO_3$溶液（$0.1mol\cdot L^{-1}$）、补钙制剂。

【实验步骤】

准确称取补钙制剂三份（每份含钙约 0.05 g），分别置于250 mL烧杯中，加入适量蒸馏水及2~5 mL 6 mol·L⁻¹ HCl溶液，并轻轻摇动烧杯，小火加热促使补钙制剂溶解。稍

Experiment 20 Determination of Calcium Content in Calcium Supplements

【Objectives】

(1) To comprehend the fundamental requirements and operations of precipitation separation.

(2) To master the principle and method of indirect determination of calcium with $KMnO_4$.

【Principle】

Some metal ions (such as alkaline earth metals, Pb^{2+}, Cd^{2+}, etc.) can react with oxalate to produce insoluble precipitates. These reactions can then be employed to determine the amount of metal ions by indirect method using $KMnO_4$. In the case of Ca^{2+}, the reactions are as follows:

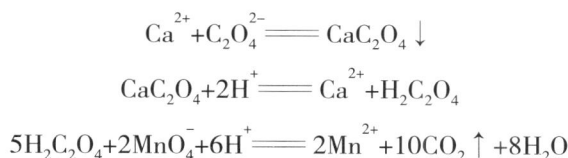

$$Ca^{2+}+C_2O_4^{2-} = CaC_2O_4 \downarrow$$

$$CaC_2O_4+2H^+ = Ca^{2+}+H_2C_2O_4$$

$$5H_2C_2O_4+2MnO_4^-+6H^+ = 2Mn^{2+}+10CO_2 \uparrow +8H_2O$$

This method can be used to determine the calcium content in some calcium supplements such as calcium gluconate, Calride, and Gaitianli, etc. The analytical results are in good agreement with the labeled quantity.

【Apparatus and Chemicals】

Apparatus: analytical balance, electric hot plate, beaker, water bath, funnel, measuring cup, acid buret, Erlenmeyer flask, wash bottle.

Chemicals: $KMnO_4$ solution (0.02 $mol \cdot L^{-1}$, prepared and standardized as in Experiment 19), $(NH_4)_2C_2O_4$ solution(0.05 $mol \cdot L^{-1}$), ammonia solution(7 $mol \cdot L^{-1}$), HCl solution(6 $mol \cdot L^{-1}$), H_2SO_4 solution (1 $mol \cdot L^{-1}$), methyl orange solution(1 $g \cdot L^{-1}$), $AgNO_3$ solution(0.1 $mol \cdot L^{-1}$), calcium supplements.

【Procedure】

Accurately weigh three portions of the calcium supplement (each containing about 0.05 g of calcium) into three 250 mL beakers respectively. Add an appropriate amount of distilled water and 2–5 mL of 6 $mol \cdot L^{-1}$ HCL solution to each beaker. Gently swirl the beaker, and heat it over low heat to facilitate dissolution. After slightly cooling, add 2–3 drops of methyl orange, and then add $NH_3 \cdot H_2O$ dropwise until the solution changes from red to yellow. While still warm, add

冷后向溶液中加入2~3滴甲基橙水溶液，再滴加氨水至溶液由红转变为黄色，趁热逐滴加入约50 mL（NH$_4$）$_2$C$_2$O$_4$溶液，在低温电热板（或水浴）上陈化30 min。

冷却后过滤（先将上层清液倾入漏斗中），将烧杯中的沉淀洗涤数次后转入漏斗中，继续洗涤沉淀至无Cl$^-$（承接洗涤液在HNO$_3$介质中以AgNO$_3$检验），将带有沉淀的滤纸铺在原烧杯的内壁上，用50 mL 1 mol·L^{-1} H$_2$SO$_4$溶液将沉淀由滤纸上洗入烧杯中，再用洗瓶洗2次，加入蒸馏水使总体积达到约100 mL，加热至70 ℃~80 ℃，用KMnO$_4$标准溶液滴定至溶液呈淡红色，再将滤纸搅入溶液中，若溶液褪色，则继续滴定，直至出现淡红色且30 s内不褪色即为终点。计算补钙制剂中钙的质量分数。

【数据记录与处理】

将数据记录到表6.5中。

表6.5   补钙制剂中钙含量的测定

项目	序号		
	I	II	III
$m_{试样}$/g			
V(KMnO$_4$)/mL			
$\omega_{钙}$/%			
$\overline{\omega}_{钙}$/%			
RAD/%			

【注意事项】

（1）实验时，必须把滤纸上的沉淀洗涤干净，滤纸一定要放入烧杯一起滴定。

（2）因为Cl$^-$与Ag$^+$的反应非常灵敏，Cl$^-$也较难洗去，故一般滤液中如无Cl$^-$，则说明杂质已洗去。

（3）洗涤沉淀时，要掌握正确的洗涤操作，注意洗涤烧杯内壁和滤纸上部，最后应确保洗净（NH$_4$）$_2$C$_2$O$_4$。应少量多次，以提高洗涤效果，避免使用过量的洗涤剂。

【思考题】

（1）加入（NH$_4$）$_2$C$_2$O$_4$时，为什么要在热溶液中逐滴加入？

（2）洗涤CaC$_2$O$_4$沉淀时，为什么要洗至无Cl$^-$？

（3）试比较KMnO$_4$法测定Ca^{2+}和配位滴定法测定Ca^{2+}的优缺点。

approximately 50 mL of $(NH_4)_2C_2O_4$ solution dropwise, and age it for 30 min on a low-temperature electric hot plate (or in a water bath).

Filter after cooling (pour the supernatant liquid into the funnel first). Wash the precipitate in the beaker several times and transfer it to the funnel. Continue washing the precipitate until no Cl^- (test the wash liquid with $AgNO_3$ in HNO_3 medium). Spread the filter paper with precipitate on the inner wall of the original beaker. Wash the precipitate from the filter paper into the beaker with 50 mL of 1 $mol \cdot L^{-1}$ H_2SO_4 solution, then wash it twice using a wash bottle. Add distilled water to make the total volume about 100 mL. Heat it to 70 ℃–80 ℃. Titrate with the $KMnO_4$ standard solution until a faint pink color appears, then stir the filter paper into the solution. If the pink color disappears, continue titrating until the faint pink color does not fade within 30 s, indicating the end of the titration. Calculate the mass fraction of calcium in the calcium supplement.

【Data Recording and Processing】

Record the experimental data in Table 6.5.

Table 6.5   Determination of Calcium Content in Calcium Supplements

Item	No.		
	I	II	III
m_{sample}/g			
$V(KMnO_4)/mL$			
$\omega(Ca)/\%$			
$\bar{\omega}(Ca)/\%$			
$RAD/\%$			

【Notes】

(1) During the experiment, the precipitate on the filter paper must be washed clean, and the filter paper must be titrated together in the beaker.

(2) Since the reaction of Cl^- and Ag^+ is very sensitive, and Cl^- is also difficult to wash away, generally, if there is no Cl^- in the filtrate, it indicates that the impurities have been washed away.

(3) When washing the precipitate, it is necessary to master the correct operation. Pay attention to washing the inner wall of the beaker and the upper part of the filter paper, and finally ensure that $(NH_4)_2C_2O_4$ is thoroughly cleaned. Wash in small amounts and multiple times to enhance the washing effect and avoid using an excessive volume of the washing agent.

【Questions】

(1) When adding $(NH_4)_2C_2O_4$, why is it necessary to add dropwise in a hot solution?

(2) When washing the CaC_2O_4 precipitate, why should it be washed to Cl^- free?

(3) Ca^{2+} can be determined by $KMnO_4$ and complexometric titrations. Compare the advantages and disadvantages of these two methods.

实验二十一   氧化还原滴定设计实验

【实验目的】

(1) 巩固理论课中学过的重要氧化还原反应的知识。

(2) 对滴定前试样的预先氧化还原处理方法和过程有一定了解。

(3) 对较复杂试样中某些组分的氧化还原滴定能设计出可行的实验方案。

【设计实验备选题】

1. 注射液中葡萄糖含量的测定

在碱性溶液中，I_2可歧化成IO^-和I^-，IO^-能定量地将葡萄糖（$C_6H_{12}O_6$）氧化成葡萄糖酸（$C_6H_{12}O_7$），过量的IO^-可进一步歧化成IO_3^-和I^-。溶液酸化后，IO_3^-又与I^-作用析出I_2，用$Na_2S_2O_3$标准溶液滴定析出的I_2，由此可计算出$C_6H_{12}O_6$的含量。

2. 福尔马林中甲醛含量的测定

福尔马林具有消毒、灭菌的作用，其主要成分是甲醛，含量在37%~40%。可采用碘量法测定甲醛含量：在碱性介质（NaOH）中，I_2发生歧化反应，生成NaIO，NaIO定量将甲醛氧化为甲酸；剩余的NaIO由于不稳定，进一步被氧化为$NaIO_3$。将溶液调整为强酸性，再加入过量的KI溶液，$NaIO_3$和KI反应生成I_2。最后以淀粉为指示剂，用$Na_2S_2O_3$标准溶液滴定析出的I_2，同时做空白实验。

3. 胱氨酸含量的测定

在酸性溶液中，BrO_3^-与Br^-发生反应生成Br_2，胱氨酸在强酸性介质中被Br_2氧化，待反应完全后，过量的Br_2可通过加入过量的KI还原，析出的I_2再用$Na_2S_2O_3$标准溶液滴定。

Experiment 21   Redox Titration
Design Experiment

[Objectives]

(1) To consolidate the knowledge of significant redox reactions learned in theoretical courses.

(2) To acquire a certain understanding of the techniques and procedures involved in pre-oxidation or pre-reduction treatment of the sample prior to titration.

(3) To be capable of designing feasible experimental protocols for the redox titration of certain components in more complex samples.

[Recommended Topics]

1. Determination of glucose content in injection

In alkaline solutions, I_2 can be disproportionated into IO^- and I^-. IO^- can quantitatively oxidize glucose ($C_6H_{12}O_6$) to gluconic acid ($C_6H_{12}O_7$), while excess IO^- further undergoes disproportionation into IO_3^- and I^-. Upon acidification, IO_3^- reacts with I^- to produce I_2, which is then titrated with a standard $Na_2S_2O_3$ solution, and finally the content of $C_6H_{12}O_6$ can be calculated.

2. Determination of formaldehyde content in formalin

Formalin possesses disinfectant and sterilizing properties, primarily attributed to its high concentration of formaldehyde (37%–40%). The quantification of formaldehyde content can be achieved through iodometry: under alkaline conditions(NaOH), I_2 undergoes disproportionation to generate NaIO, which subsequently oxidizes formaldehyde into formic acid in a quantitative manner. Due to its inherent instability, the remaining NaIO is further converted into $NaIO_3$ via oxidation. The solution is then adjusted to strong acidity before excess KI solution is introduced. $NaIO_3$ reacts with KI, resulting in the formation of I_2. Ultimately, the I_2 produced is titrated using $Na_2S_2O_3$ standard solution with starch as an indicator; a blank experiment is conducted for reference.

3. Determination of cystine content

The bromate ion (BrO_3^-) undergoes a reaction with bromide ion (Br^-) to produce molecular bromine (Br_2) in an acidic solution, while cystine is oxidized by Br_2 in a strongly acidic medium. Once the reaction is complete, any excess Br_2 can be reduced by adding an excess of potassium

4. 食用油过氧化值的测定

油脂被氧化生成过氧化物的多少常以过氧化值来表示。油脂的过氧化值是指100 g油脂中所含的过氧化物在酸性环境下与KI作用时析出I_2的克数，采用$Na_2S_2O_3$标准溶液滴定。

5. 水中的溶解氧（DO）的测定

水中的溶解氧在碱性介质中可将$Mn(OH)_2$氧化为棕色的$MnO(OH)_2$，后者在酸性介质中溶解并能与I^-定量作用产生I_2，析出来的I_2则可用$Na_2S_2O_3$标准溶液滴定。

6. HCOOH与HAc混合液中各组分含量的测定

以酚酞为指示剂，用NaOH标准溶液滴定总酸量，在强碱性介质中向试样溶液加入过量$KMnO_4$标准溶液，此时甲酸被氧化为CO_2，MnO_4^-还原为MnO_4^{2-}，并歧化生成MnO_4^-及MnO_2。加酸，加入过量的KI还原过量部分的MnO_4^-及歧化生成的MnO_4^-和MnO_2至Mn^{2+}，再用$Na_2S_2O_3$标准溶液滴定析出的I_2。

7. 药用硫酸亚铁的测定

在硫酸酸性溶液中，$KMnO_4$能将亚铁氧化成三价铁，利用$KMnO_4$自身作指示剂指示滴定终点。

8. 化学需氧量的测定

化学需氧量（Chemical Oxygen Demand，COD）是指在特定条件下，用强氧化剂处理水样时，水样所消耗的氧化剂的量，常用每升水消耗O_2的量来表示（$mg \cdot L^{-1}$）。

重铬酸钾法：在强酸性条件下，向水样中加入过量的$K_2Cr_2O_7$，让其与试样中的还原性物质充分反应，剩余的$K_2Cr_2O_7$以邻二氮菲为指示剂，用硫酸亚铁铵标准溶液返滴定。根据消耗的$K_2Cr_2O_7$溶液的体积和浓度，计算水样的需氧量。氯离子会干扰测定，可加硫酸银除去。该法适用于工业污水及生活污水等含有较多复杂污染物的水样的测定。

酸性$KMnO_4$法：在酸性条件下，向水样中加入过量的$KMnO_4$溶液，并加热溶液让其充分反应，然后再向溶液中加入过量的$Na_2C_2O_4$标准溶液还原多余的$KMnO_4$，剩余的$Na_2C_2O_4$再用$KMnO_4$溶液返滴定。根据$KMnO_4$的浓度和水样所消耗的$KMnO_4$溶液体积，计

iodide（KI）, and the produced iodine（I_2）can be titrated using a standard solution of sodium thiosulfate（$Na_2S_2O_3$）.

4. Determination of peroxide value in edible oils

The peroxide value is commonly used to quantify the amount of peroxides generated through oil oxidation. The peroxide value of grease refers to the grams of iodine（I_2）produced when 100 g of grease, containing peroxide, reacts with KI in an acidic environment and is subsequently titrated using a standard $Na_2S_2O_3$ solution.

5. Determination of dissolved oxygen（DO）in water

The dissolved oxygen in water can oxidize $Mn(OH)_2$ to brown $MnO(OH)_2$ in an alkaline medium. This compound can be solubilized in an acidic medium and react quantitatively with I^- to generate I_2, which can then be titrated using a standard solution of $Na_2S_2O_3$.

6. Determination of the contents of individual component in a mixed solution of formic acid（HCOOH）and acetic acid（HAc）

Using phenolphthalein as an indicator, the total acid content is titrated with a standard solution of NaOH. An excess of $KMnO_4$ standard solution is added to the sample solution in a strongly alkaline medium. In this case, formic acid undergoes oxidation to CO_2, MnO_4^- is reduced to MnO_4^{2-}, which then disproportionates into MnO_4^- and MnO_2. After adding acid, the excess KI is introduced to reduce the surplus portion of MnO_4^- as well as the disproportionated products（MnO_4^- and MnO_2）into Mn^{2+}, followed by titration of I_2 produced with a standard solution of $Na_2S_2O_3$.

7. Determination of ferrous sulfate for medicinal purposes

In a sulfuric acid solution, the oxidation of ferrous iron to ferric iron can be achieved by employing $KMnO_4$ as an oxidizing agent and utilizing $KMnO_4$ itself as an indicator to signify the endpoint of titration.

8. Determination of chemical oxygen demand

Chemical Oxygen Demand（COD）refers to the quantity of oxidizing agent consumed by a water sample under specific conditions when treated with a strong oxidizing agent, typically expressed as the amount of oxygen（O_2）consumed per liter of water（$mg \cdot L^{-1}$）.

The potassium dichromate method involves adding excess $K_2Cr_2O_7$ to the water sample under strongly acidic condition to fully react with the reducing substances, followed by back titration of the remaining $K_2Cr_2O_7$ with ammonium ferrous sulfate standard solution using o-phenanthroline as an indicator. The oxygen demand of the water sample is then calculated based on the volume and concentration of the consumed $K_2Cr_2O_7$ solution. Interference from chloride ions can be eliminated by adding silver sulfate. This method is suitable for analyzing water samples containing complex pollutants such as industrial or domestic sewage.

In the acid $KMnO_4$ method, an excess of $KMnO_4$ solution is added to the water sample

算水样的需氧量。该法适用于污染不十分严重的地面水和河水等的化学需氧量的测定。

若水样中Cl⁻含量较高，可加入Ag_2SO_4消除干扰，也可改用碱性$KMnO_4$法进行测定。

under acidic conditions. The solution is then heated to ensure complete reaction. Next, excess $Na_2C_2O_4$ standard solution is introduced to reduce the remaining $KMnO_4$. Finally, the remaining $Na_2C_2O_4$ is back titrated with the $KMnO_4$ solution. By considering the concentration and volume of $KMnO_4$ consumed by the water sample, the oxygen demand can be calculated accurately. This technique is suitable for determining chemical oxygen demand in surface water and lightly polluted river water. If there is a high Cl^- content in the water sample, Ag_2SO_4 can be added to eliminate the interference or alternatively, the alkaline $KMnO_4$ method can be employed.

第七章　分光光度法

Chapter 7　Spectrophotometry

实验二十二　邻二氮菲分光光度法测定铁的条件实验及配合物组成的测定

【实验目的】

（1）掌握 722S 型分光光度计的使用方法，并明晰其工作原理。

（2）熟练掌握比色皿的正确使用方式。

（3）学习绘制吸收曲线的方法。

（4）学习怎样选择吸光光度分析的实验条件。

【实验原理】

在运用可见光吸光光度法进行测定时，倘若被测组分自身颜色较浅，抑或无色，则需加入显色剂与其发生反应生成有色化合物，方可进行测定。显色反应会受到多种因素的影响，例如，溶液的酸度、显色剂的用量、有色溶液的稳定性、温度、溶剂以及干扰物质等。通常通过条件实验来确定上述各因素的最佳条件，条件试验的简便方法为：改变某一实验条件，固定其余条件，测量得到一系列的吸光度值，绘制吸光度与某实验条件的关系曲线，依据曲线来确定该实验条件的适宜值。

本实验将借由邻二氮菲-Fe^{2+}显色反应来探究如何确定一个光度分析方法的实验条件。

【仪器与试剂】

仪器：分光光度计、比色皿或常量瓶（50 mL × 8）。

试剂：

（1）铁标准溶液（1.00×10^{-3} mol·L^{-1}，0.5 mol·L^{-1} HCl溶液）：准确称取0.482 2 g $NH_4Fe(SO_4)_2 \cdot 12H_2O$，置于烧杯中，加入80 mL 6 mol·$L^{-1}$ HCl溶液和适量蒸馏水，溶解后转移至1 L容量瓶中，用蒸馏水稀释至刻度，摇匀。

（2）邻二氮菲溶液（1.5 g·L^{-1}）、盐酸羟胺水溶液（100 g·L^{-1}）、NaAc溶液（1 mol·L^{-1}）、NaOH溶液（1 mol·L^{-1}）。

Experiment 22   Conditional Test for the Determination of Iron by 1,10–Phenanthroline Spectrophotometry and the Determination of Complex Composition

【Objectives】

(1) To acquire the skills of operating the 722S spectrophotometer and comprehend its working principle.

(2) To skillfully master the correct way of using cuvettes.

(3) To study the method of plotting absorption curves.

(4) To learn how to select the experimental conditions for spectrophotometric analysis.

【Principles】

When conducting measurements using visible light absorption photometry, if the component under test has a pale color or is colorless, a chromophoric reagent needs to be added to react with it to form a colored derivative for the measurement. The chromophoric reaction is influenced by various factors, such as the acidity of the solution, the dosage of the chromophoric reagent, the stability of the colored solution, temperature, solvent, and interfering substances. Generally, the optimal conditions for each of these factors are determined through experiments. A straightforward approach for condition tests is to vary one experimental condition while keeping the others fixed, measure a series of absorbance values, draw a relationship curve of absorbance versus the varied experimental condition, and determine the appropriate value of the experimental condition based on the curve.

This experiment will explore how to determine the experimental conditions of a photometric analysis through the 1,10–phenanthroline–Fe^{2+} chromophoric reaction.

【Apparatus and Chemicals】

Apparatus: spectrophotometer, colorimetric tubes or volumetric flasks (50 mL × 8).

Chemicals:

(1) Iron standard solution (1.00×10^{-3} mol·L^{-1}, 0.5 mol·L^{-1} HCl solution): Weigh accurately 0.482 2 g of $NH_4Fe(SO_4)_2 \cdot 12H_2O$ and place it in a beaker. Add 80 mL of 6 mol·L^{-1} HCl solution and an appropriate amount of distilled water. After dissolution, transfer it to a 1 L volumetric flask. Dilute it to the mark with distilled water and shake well.

(2) 1,10-Phenanthroline solution (1.5 g·L^{-1}), hydroxylamine hydrochloride aqueous

【实验步骤】

1. 吸收曲线的绘制与测量波长的选择

用吸量管吸取2 mL 1.0×10^{-3} mol·L^{-1} 铁标准溶液，注入50 mL比色管中，加入1 mL 100 g·L^{-1} 盐酸羟胺水溶液，摇匀（原则上每加入一种试剂后均需摇匀），加入2 mL 1.5g·L^{-1} 邻二氮菲溶液、5 mL 1mol·L^{-1} NaAc溶液，以蒸馏水稀释至刻度，摇匀。放置 10 min后，在光度计上用1 cm比色皿，采用蒸馏水作为参比，于450～550 nm之间，每隔 10 nm测量一次吸光度；在最大吸收波长附近，每隔2 nm测量一次，以波长为横坐标，吸光度为纵坐标，绘制$A-\lambda$吸收曲线，选择测量的适宜波长，通常选用最大吸收波长λ_{max}。

2. 显色剂用量的选择

取7支50 mL比色管，各加入2 mL的1.0×10^{-3} mol·L^{-1} 铁标准溶液以及1 mL 100 g·L^{-1}盐酸羟胺水溶液，摇匀。分别添加0.30 mL、0.50 mL、0.80 mL、1.00 mL、1.50 mL、2.00 mL 及4.00 mL 1.5g·L^{-1} 邻二氮菲溶液，5.0 mL 1 mol·L^{-1} NaAc溶液，用蒸馏水稀释至刻度，摇匀。放置10 min后，在光度计上用1 cm 比色皿，选择适宜（由步骤1所选定的）波长，以蒸馏水作为参比，测定吸光度。以加入的邻二氮菲体积作为横坐标，相应的吸光度作为纵坐标，绘制吸光度-显色剂用量（$A-V_{显色剂}$）曲线，从而确定最佳显色剂用量。

3. 溶液pH值的影响

取8支50 mL比色管，每支加入2 mL的1.00×10^{-3} mol·L^{-1} 铁标准溶液以及1 mL 100 g·L^{-1}盐酸羟胺水溶液，摇匀后加入2 mL 1.5g·L^{-1} 邻二氮菲溶液，接着再用5 mL 吸量管分别加入0 mL、0.2 mL、0.5 mL、1 mL、1.5 mL、2 mL、2.5 mL、3 mL 1 mol·L^{-1}的NaOH溶液，用蒸馏水稀释至刻度，摇匀。放置10 min，于选定的波长，以蒸馏水作为参比，用1 cm 吸收池测量各溶液的吸光度。用精密pH试纸（或pH计）测定各溶液pH。绘制$A-pH$曲线，确定适宜的pH范围。

4. 显色时间的影响

取一支50 mL的比色管，加入2 mL的1.00×10^{-3} mol·L^{-1} 铁标准溶液、1 mL 100 g·L^{-1}盐酸羟胺水溶液，摇匀。接着加入2 mL 1.5 g·L^{-1} 邻二氮菲溶液，5 mL 1 mol·L^{-1} NaAc溶液，用蒸馏水稀释至刻度，摇匀。即刻在选定的波长下，用1 cm 比色皿，以蒸馏水作为参比，测定吸光度。随后测量放置5 min、10 min、15 min、20 min、30 min、60 min、120 min 时相应的吸光度。以时间t作为横坐标，吸光度A作为纵坐标，在坐标纸上绘制$A-t$曲线，从曲线上观察显色反应完全所需的时间及其稳定性，并确定适宜的测量时间。

solution $(100 \text{ g} \cdot \text{L}^{-1})$, NaAc solution $(1 \text{ mol} \cdot \text{L}^{-1})$, NaOH solution $(1 \text{ mol} \cdot \text{L}^{-1})$.

[Procedure]

1. Construction of absorption curve and selection of measuring wavelength

Pipet 2 mL of $1.0 \times 10^{-3} \text{ mol} \cdot \text{L}^{-1}$ iron standard solution into a 50 mL colorimetric tube. Add 1 mL of $100 \text{ g} \cdot \text{L}^{-1}$ hydroxylamine hydrochloride aqueous solution, shake well (in principle, shaking is required after adding each reagent), add 2 mL of $1.5 \text{ g} \cdot \text{L}^{-1}$ 1, 10-phenanthroline solution and 5 mL of $1 \text{ mol} \cdot \text{L}^{-1}$ NaAc solution, dilute to the mark with distilled water, and shake well. After 10 min of standing, use a 1 cm cuvette, with distilled water as the reference, measure the absorbance between 450–550 nm at intervals of 10 nm; near the maximum absorption wavelength, measure at intervals of 2 nm. Take the wavelength as the abscissa and the absorbance as the ordinate, plot an $A-\lambda$ absorption curve, and select the appropriate wavelength for measurement, usually the maximum absorption wavelength λ_{max}.

2. Selection of the amount of chromophoric reagent

Take seven 50 mL colorimetric tubes. Add 2 mL of $1.0 \times 10^{-3} \text{ mol} \cdot \text{L}^{-1}$ iron standard solution and 1 mL of $100 \text{ g} \cdot \text{L}^{-1}$ hydroxylamine hydrochloride aqueous solution to each tube. Shake well. Respectively add 0.30 mL, 0.50 mL, 0.80 mL, 1.00 mL, 1.50 mL, 2.00 mL, and 4.00 mL of $1.5 \text{ g} \cdot \text{L}^{-1}$ 1,10-phenanthroline solution, 5 mL of $1 \text{ mol} \cdot \text{L}^{-1}$ NaAc solution, and dilute to the mark with distilled water. Shake well. After 10 min of standing, use a 1 cm cuvette, select the appropriate wavelength (determined in step 1), and use distilled water as the reference to measure the absorbance of each solution. Take the volume of 1,10-phenanthroline added as the abscissa and the corresponding absorbance as the ordinate, plot an absorbance-amount of chromophoric reagent $(A-V_{\text{chromogenic reagent}})$ curve, thereby determining the optimal amount of chromophoric reagent.

3. Influence of solution pH

Take eight 50 mL colorimetric tubes. Add 2 mL of $1.00 \times 10^{-3} \text{ mol} \cdot \text{L}^{-1}$ iron standard solution and 1 mL of $100 \text{ g} \cdot \text{L}^{-1}$ hydroxylamine hydrochloride aqueous solution to each tube. Shake well and then add 2 mL of $1.5 \text{ g} \cdot \text{L}^{-1}$ 1,10-phenanthroline solution. Subsequently, add 0 mL, 0.2 mL, 0.5 mL, 1 mL, 1.5 mL, 2 mL, 2.5 mL, and 3 mL of $1 \text{ mol} \cdot \text{L}^{-1}$ NaOH solution respectively using a 5 mL pipet. Dilute to the mark with distilled water and shake well. Let it stand for 10 min. At the selected wavelength, using distilled water as the reference, measure the absorbance of each solution with a 1 cm cell. Determine the pH of each solution using a precise pH test paper (or pH meter). Plot an $A-$pH curve to determine the appropriate pH range.

4. Influence of reaction time

Take a 50 mL colorimetric tube, add 2 mL of $1.00 \times 10^{-3} \text{ mol} \cdot \text{L}^{-1}$ iron standard solution and 1 mL of $100 \text{ g} \cdot \text{L}^{-1}$ hydroxylamine hydrochloride aqueous solution. Shake well. Subsequently,

5. 邻二氮菲与铁的配合比测定（摩尔比法）

取8支50 mL 比色管，分别向其中加入1 mL 的1.00×10^{-3} mol·L^{-1}铁标准溶液，1 mL 100 g·L^{-1}盐酸羟胺溶液，摇匀。依次加入1.5 g·L^{-1}邻二氮菲溶液0.5 mL、1 mL、1.5 mL、2 mL、2.5 mL、3 mL、4 mL、4.5 mL，再分别加入5 mL 1 mol·L^{-1} NaAc 溶液，用蒸馏水稀释至刻度，摇匀。放置10 min后，在选用的波长下，用1 cm 比色皿，以蒸馏水作为参比，测定吸光度。以邻二氮菲与铁的浓度比$c(\text{Phen})/c(\text{Fe})$作为横坐标，吸光度$A$作为纵坐标作图，根据曲线上前后两部分延长线的交点位置，确定铁与邻二氮菲的络合比。

【数据记录与处理】

将实验数据记录至表7.1至表7.5中。

表7.1   吸收曲线的绘制

波长λ/nm	
吸光度A	

表7.2   显色剂用量

项目	序号							
	1	2	3	4	5	6	7	8
$V_{显色剂}$/mL								
吸光度A								

表7.3   溶液酸度的影响

项目	序号							
	1	2	3	4	5	6	7	8
$V(\text{NaOH})$/mL								
pH值								
吸光度A								

add 2 mL of $1.5 \text{ g} \cdot \text{L}^{-1}$ 1, 10–phenanthroline solution, 5 mL of $1 \text{ mol} \cdot \text{L}^{-1}$ NaAC solution, and dilute to the mark with distilled water. Shake well. Immediately, at the selected wavelength, using a 1 cm cuvette and distilled water as the reference, measure the absorbance. Subsequently, measure the corresponding absorbances after 5 min, 10 min, 15 min, 20 min, 30 min, 60 min, and 120 min. With time t as the abscissa and absorbance A as the ordinate, draw an $A-t$ curve on the coordinate paper. Observe from the curve the time required for the chromophoric reaction to be complete and its stability, and determine the appropriate measurement time.

5. Determination of ligand–to–metal ratio between 1,10-phenanthroline and iron (molar ratio method)

Take eight 50 mL colorimetric tubes and add 1 mL of $1.00 \times 10^{-3} \text{ mol} \cdot \text{L}^{-1}$ iron standard solution to each one. Add 1 mL of $100 \text{ g} \cdot \text{L}^{-1}$ hydroxylamine hydrochloride solution and shake well. Sequentially add 0.5 mL, 1 mL, 1.5 mL, 2 mL, 2.5 mL, 3 mL, 4 mL, and 4.5 mL of $1.5 \text{ g} \cdot \text{L}^{-1}$ 1,10-phenanthroline solution, and then add 5 mL of $1 \text{ mol} \cdot \text{L}^{-1}$ NaAc solution to each tube. Dilute to the mark with distilled water and shake well. After standing for 10 min, using a 1 cm cuvette and distilled water as the reference, measure the absorbance at the selected wavelength. Plot a graph with the concentration ratio of 1,10-phenanthroline to iron, $c(\text{Phen})/c(\text{Fe})$, as the abscissa and the absorbance A as the ordinate. Determine the composition ratio of iron to 1,10-phenanthroline based on the intersection point of the extended lines of the two parts of the curve.

【Data Recording and Processing】

Record the experimental data in Table 7.1 to Table 7.5.

Table 7.1   Absorption Curve

Wavelength(λ)/nm	
Absorbance(A)	

Table 7.2   Amount of Chromophoric Reagent

Item	No.							
	1	2	3	4	5	6	7	8
$V_{\text{chromophoric reagent}}$/mL								
Absorbance(A)								

Table 7.3   Influence of Solution Acidity

Item	No.							
	1	2	3	4	5	6	7	8
$V(\text{NaOH})$/mL								
pH value								
Absorbance(A)								

表7.4   显色时间的影响

时间t/min	
吸光度A	

表7.5   邻二氮菲与铁的配合比测定

项目	序号							
	1	2	3	4	5	6	7	8
$c(\mathrm{Phen})/(\mathrm{mol \cdot L^{-1}})$								
$c(\mathrm{Phen})/c(\mathrm{Fe})$								
吸光度A								

【注意事项】

(1) 各类试剂的加入顺序不得颠倒。

(2) 同一组溶液必须在同一台仪器上予以测量。

(3) 分光光度计需预热 30 min，待其稳定后方可进行测量。

(4) 每变更一个波长，需先用参比溶液将透光度调至 100%。

【思考题】

(1) 以邻二氮菲测定铁时，盐酸羟胺发挥何种作用？若要测定混合铁中的亚铁含量，是否需要加入盐酸羟胺？

(2) 使用分光光度计时应注意哪些方面？

(3) 采用邻二氮菲分光光度法测量铁时，主要需要控制哪些反应条件？

Table 7.4   Influence of Reaction Time

t/min	
Absorbance(A)	

Table 7.5   Stoichiometric Ratio Between 1, 10-Phenanthroline and Iron

Iten	No.							
	1	2	3	4	5	6	7	8
$c(\text{Phen})/(\text{mol}\cdot\text{L}^{-1})$								
$c(\text{Phen})/c(\text{Fe})$								
Absorbance(A)								

【Notes】

(1) The addition sequence of various reagents shall not be reversed.

(2) The solutions of the same group must be measured on the same instrument.

(3) The spectrophotometer needs to be preheated for 30 min and allowed to stabilize before measurement.

(4) Whenever changing a wavelength, the transmittance should be adjusted to 100% with the reference solution first.

【Questions】

(1) Whendetermining ironwith 1, 10-phenanthroline, what role does hydroxylamine hydrochloride play? If determining the ferrous content in a mixed iron sample, is it necessary to add hydroxylamine hydrochloride?

(2) What aspects should be noted when using a spectrophotometer?

(3) When measuring iron by the 1, 10-phenanthro line spectrophotometric method, what reaction conditions mainly need to be controlled?

实验二十三　邻二氮菲分光光度法测定微量铁

【实验目的】

(1) 掌握分光光度计的使用方法。

(2) 掌握邻二氮菲分光光度法测定铁的原理和方法。

(3) 掌握利用标准曲线进行微量组分测定的方法。

【实验原理】

邻二氮菲是铁的一种优良的显色剂，在pH 2～9的溶液中，Fe^{2+}能与其生成1:3的橙红色配合物（$\lg \beta_3 = 21.3$）[①]。最大吸收波长510 nm处的摩尔吸光系数为1.1×10^4 $L \cdot mol^{-1} \cdot cm^{-1}$。在一定浓度范围内，$Fe^{2+}$的浓度与配合物吸光度的关系遵循朗伯–比尔定律。有关反应式如下：

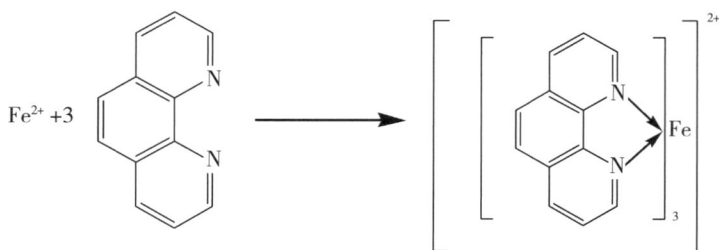

该显色反应选择性很高，形成的配合物较稳定，在有还原剂的情况下，颜色可保持数月不变。由于Fe^{3+}也可与邻二氮菲生成1:3的淡蓝色配合物，$\lg \beta_3 = 14.1$，因此，在显色反应前，需加入盐酸羟胺将Fe^{3+}全部还原成Fe^{2+}。

---

①该配合物的累积稳定常数为β_3，其对数值大表示其性质非常稳定，故可用于定量测定Fe^{2+}的浓度。

Experiment 23   Spectrophotometric Determination of Trace Iron with 1,10-Phenanthroline

【Objectives】

(1) To master the use of the spectrophotometer.

(2) To grasp the principle and method in spectrophotometric determination of iron with 1,10-phenanthroline.

(3) To master the determination of trace components using the standard curve.

【Principle】

The red-orange 1 : 3 complex that forms between Fe^{2+} and 1,10-phenanthroline is useful in determining iron in water supplies. Fe^{2+} is quantitatively complexed in the pH range between 2 and 9 $(lg\beta_3=21.3)$[①]. The molar absorptivity(ε) at the maximum absorption wavelength(510 nm) is 1.1×10^4 $L \cdot mol^{-1} \cdot cm^{-1}$. The relationship between the concentration of Fe^{2+} and the absorbance of the complex follows Beer's law in a certain concentration range. Complex formation is thus described by the equation:

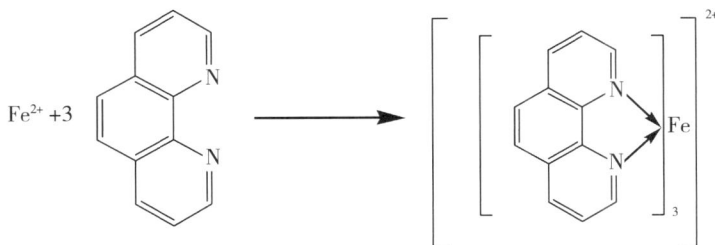

This chromophoric reaction exhibits exceptional selectivity, forming stable complexes that maintain chromatic stability for months with reducing agents. To prevent interference from Fe^{3+} to 1,10-phenanthroline coordination(1 : 3 pale blue complex), quantitative reduction to Fe^{2+} using hydroxylamine hydrochloride must precede analysis.

---

① The complex's overall formation constant $(lg\beta_3)$ indicates high stability, enabling quantitative determination of Fe^{2+} concentration.

【仪器与试剂】

仪器：分光光度计、容量瓶或比色管（50 mL×7）。

试剂：

（1）100 $\mu g \cdot mL^{-1}$ 铁标准溶液：准确称取0.863 4 g $NH_4Fe(SO_4)_2 \cdot 12H_2O$，置于烧杯中，用20 mL 6 $mol \cdot L^{-1}$ HCl溶液和适量蒸馏水溶解后，定量转移到1 L容量瓶中，用蒸馏水稀释至刻度，摇匀。

（2）100 $g \cdot L^{-1}$ 盐酸羟胺水溶液，1.5 $g \cdot L^{-1}$ 邻二氮菲溶液，1 $mol \cdot L^{-1}$ NaAc溶液，6 $mol \cdot L^{-1}$ HCl溶液。

【实验步骤】

1. 铁标准溶液（10 $\mu g \cdot mL^{-1}$）的配制

移取100 $\mu g \cdot mL^{-1}$ 的铁标准溶液10 mL，置于100 mL容量瓶中，加入2 mL 6 $mol \cdot L^{-1}$ HCl溶液，用蒸馏水稀释至刻度，摇匀。

2. 标准系列溶液的配制

移取步骤1所配制的铁标准溶液（10 $\mu g \cdot mL^{-1}$）：将0 mL、2 mL、4 mL、6 mL、8 mL、10 mL铁标准溶液依次放入6个洁净的50 mL容量瓶（编号0~5）中，分别加入100 $g \cdot L^{-1}$ 盐酸羟胺溶液1 mL，稍摇动（原则上每加入一种试剂都要摇匀），再加入1.5 $g \cdot L^{-1}$ 邻二氮菲溶液2 mL及1 $mol \cdot L^{-1}$ NaAc溶液5 mL，用蒸馏水稀释至刻度，充分摇匀。

3. 未知样的配制

准确移取5 mL待测样品于50 mL容量瓶（编号6）中，加入100 $g \cdot L^{-1}$ 盐酸羟胺溶液1 mL，稍摇动，再加入1.5 $g \cdot L^{-1}$ 邻二氮菲溶液2 mL及1 $mol \cdot L^{-1}$ NaAc溶液5 mL，用蒸馏水稀释至刻度，充分摇匀。

4. 吸收光谱的测定和最大吸收波长的确定

如已做实验二十二实验步骤1，可略去这步。

3号铁标准溶液放置10 min后，在分光光度计上，用1 cm比色皿，以容量瓶0号溶液为参比，在450～550 nm之间，每隔10 nm测定一次3号铁标准溶液的吸光度A，在最大吸收波长附近，每隔2 nm测定一次吸光度。以波长为横坐标，吸光度为纵坐标，绘制$A-\lambda$吸收曲线，从而测定铁的最大吸收波长λ_{max}。根据3号铁标准溶液的浓度和比色皿的厚度，计算铁（Ⅱ）配合物在最大吸收波长处的摩尔吸光系数。

【Apparatus and Chemicals】

Apparatus: Spectrophotometer, volumetric flasks or colorimetric tubes (50 mL × 7).

Chemicals:

(1) 100 $\mu g \cdot mL^{-1}$ iron standard solution: Accurately weigh 0.863 4 g of $NH_4Fe(SO_4)_2 \cdot 12H_2O$ and place it in a beaker. Dissolve it with 20 mL of 6 $mol \cdot L^{-1}$ HCl solution and an adequate amount of distilled water. Subsequently, quantitatively transfer the solution to a 1 L volumetric flask. Dilute it to the mark with distilled water. Stopper and invert several times to mix.

(2) 100 $g \cdot L^{-1}$ aqueous solution of hydroxylamine hydrochloride, 1.5 $g \cdot L^{-1}$ aqueous solution of 1, 10-phenanthroline(Phen), 1 $mol \cdot L^{-1}$ sodium acetate(NaAc)solution, 6 $mol \cdot L^{-1}$ hydrochloric acid (HCl) solution.

【Procedure】

1. Preparation of iron standard solutions (10 $\mu g \cdot mL^{-1}$)

Pipet 10 mL of 100 $\mu g \cdot mL^{-1}$ iron standard solution into a 100 mL volumetric flask. Add 2 mL of 6 $mol \cdot L^{-1}$ HCl and fill to the mark with distilled water. Stopper and invert several times to mix.

2. Preparation of the standard series

Into 6 clean 50 mL volumetric flasks (marked 0–5), add with pipets 0 mL, 2 mL, 4 mL, 6 mL, 8 mL, 10 mL of the 10 $\mu g \cdot mL^{-1}$ iron standard solution. To each of the flasks, add 1 mL of the hydroxylamine hydrochloride solution, swirl for mixing (it is generally advisable to mix well after each addition of a reagent), 2 mL of Phen, 5 mL of NaAc solution. Add distilled water to the mark. Stopper and invert several times to mix.

3. Preparation of the sample solution

Pipet 5 mL of the unknown to a 50 mL volumetric flask (marked 6). Sequentially add 1 mL of the hydroxylamine hydrochloride solution, swirl it, then add 2 mL of Phen and 5 mL of NaAc solution. Dilute to the mark with distilled water and mix thoroughly.

4. Measurement of absorption spectrum and determination of the maximum absorption wavelength

This step can be omitted if Step 1 of Experiment 22 has been conducted.

With 1 cm cells, use the blank solution (No.0) as the reference and obtain the absorption spectrum of the No.3 solution by measuring the absorbance from 450–550 nm. Take readings at 10 nm intervals. Near the vicinity of the absorption maximum, take readings at 2 nm intervals. Plot the absorbance against the wavelength and select the wavelength of the absorption maximum (λ_{max}). Calculate the molar absorptivity of the iron (Ⅱ) complex at the λ_{max}, based on the concentration of the No.3 solution and the thickness of the cuvette.

5. 标准曲线的绘制

放置10 min后，以不加铁标准溶液的0号试液为参比，以选定的最大吸收波长λ_{max}为测定波长，依次测定标准系列溶液（1~5）中各溶液的吸光度A值。以铁的质量浓度为横坐标，A值为纵坐标，绘制标准曲线。

6.未知样品中铁含量的测定

用测量标准曲线同样的方法测定未知样品（6）的吸光度A值。在标准曲线上查得含铁浓度，或通过最小二乘法拟合标准曲线，利用Excel绘制求出曲线的斜率和截距，结合未知样的吸光度计算含铁浓度，最后计算未知样品中铁的总含量。

【数据记录与处理】

将实验数据记录至表7.6、表7.7中。

表7.6　不同波长下3号标准溶液的吸光度

波长λ/nm	
吸光度A	

绘制A–λ曲线（请用方格纸或Excel画图），确定最大吸收波长为_____nm。

表7.7　标准曲线的绘制及未知样测定

项目	序号						
	0	1	2	3	4	5	6（未知）
$V(Fe)$/ mL	0	2	4	6	8	10	5
$c(Fe^{2+})$/（$\mu g \cdot mL^{-1}$）							
吸光度A							

根据表7.7数据绘制标准曲线（请用方格纸或Excel画图）。

其回归方程为：_____，R^2=_____。

由标准曲线和未知样吸光度可计算得未知铁浓度为c = _____ $\mu g \cdot mL^{-1}$。

计算得摩尔吸光系数 ε = _____。

【注意事项】

（1）不能颠倒各种试剂的加入顺序。

（2）选择好最佳波长后不要再改变。

5. Plotting the standard curve

Determine the absorbance of the standards (1–5) with respect to the blank (0) at the wavelength of maximum absorbance. Prepare a standard curve by plotting the absorbance of the standards against concentration.

6. Determination of iron in the unknown sample

Measure the absorbance of the sample (6) with respect to the blank (0). From the standard curve and the unknown's absorbance, determine the final concentration of iron in your unknown solution. Perform the calculations of the measured concentration by entering the formula in a cell of the spreadsheet using the slope and intercept values and the measured absorbance.

【Data Recording and Processing】

Record the experimental data in Table 7.6 and Table 7.7.

Table 7.6　Absorbance of Standard Solution No.3 at Various Wavelengths

Wavelength λ/nm	
Absorbance (A)	

Plot $A-\lambda$ curve (please use graph paper or Excel for graphing) and determine the maximum absorption wavelength as_____nm.

Table 7.7　Plotting of the Standard Curve and Determination of the Unknown

Item	No.						
	0	1	2	3	4	5	6 (unknown)
$V(Fe)$/ mL	0	2	4	6	8	10	5
$c(Fe^{2+})$/($\mu g \cdot mL^{-1}$)							
Absorbance (A)							

The standard curve is plotted based on the data in Table 7.7 (please use graph paper or Excel for graphing).

Its regression equation is: _____, $R^2 =$_____.

From the standard curve and the absorbance of the unknown sample, the unknown iron concentration can be calculated as $c =$ _____ $\mu g \cdot mL^{-1}$.

The molar absorptivity is calculated as $\varepsilon =$ _____.

【Notes】

(1) The addition sequence of various reagents must not be reversed.

(2) The selected optimal wavelength should not be altered.

(3) The determination of iron content in the sample and the generation of the standard curve can be conducted concurrently.

(4) With each change of wavelength, the transmittance should first be adjusted to 100%

(3) 试样中铁含量的测定和标准曲线的绘制可同时进行。

(4) 每改变一次波长，要先用参比溶液调节透光度至100%。

【思考题】

(1) 什么是吸收曲线？什么是标准曲线？实验中测定这两个曲线的目的是什么？

(2) 参比溶液的作用是什么？在本实验中可否用蒸馏水作参比溶液？

using the reference （blank） solution.

【Questions】

（1） What is an absorption curve? What is a standard curve? What is the purpose of determining these two curves in the experiment?

（2） What is the role of the reference solution? Can distilled water serve as the reference solution in this experiment?

实验二十四　研究性实验：
光纤光谱法测定 pH 值

　　与传统的电化学分析相比，光纤化学传感器具有一系列突出的优点，如体积小、不带电、抗电磁干扰性能强、无污染等，还可进行远程测量和多路检测等，在各类化学反应、环境监测、生物医学领域，及地下矿井、武器试验等危险场合的遥测遥控方面都有广阔的应用前景，受到国际上的广泛重视。

　　图7.1是光吸收型光纤化学pH传感器的基本工作原理图。光源发射的光由入射光纤传导至探头（调制区），涂覆在光纤探头上的敏感膜与目标分子发生作用时，引起了光吸收的变化，通过这种变化可以检测目标分子的浓度或其他化学性质。此类仪器的特点是不需要吸收池，直接将探头插入试样溶液中，即可在原位进行测定，不受外界光线的影响。

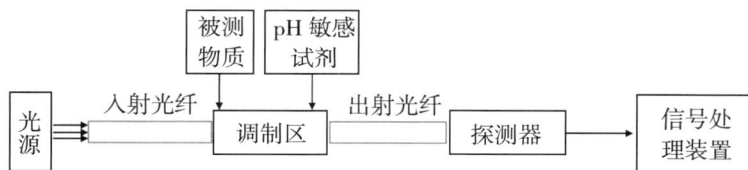

图7.1　吸收型光纤化学pH传感器基本工作原理

　　吸收型光纤化学pH传感器融合了化学制膜、光纤技术以及分光光度法等技术。传感器探针是光纤化学pH传感器中最重要的组成元件，它是一个试剂固定装置，可以将敏感膜（膜成分一般由pH敏感指示剂、固定试剂支持剂、增塑剂等组成）固定于光纤末端，也可以将敏感膜直接涂敷在腐蚀掉包层的光纤纤芯上，构成传感器探针。

Experiment 24　Research Experiment:
Measurement of pH Value by Fiber Optic
Spectroscopy

In comparison with traditional electrochemical analysis, fiber optic chemical sensors possess a series of prominent advantages, such as small size, no electrical charge, strong anti-electromagnetic interference performance, no pollution, etc. They can also conduct remote measurement and multi-channel detection. They have broad application prospects in various chemical reactions, environmental monitoring, biomedical fields, as well as in remote sensing and control in dangerous situations such as underground mines and weapon tests, and have thus received extensive attention internationally.

Figure 7.1 depicts the working principle of the absorption type fiber optic chemical pH sensor. The light emitted by the source is conducted through the incident fiber to the probe (modulation region). When the sensitive film coated on the fiber optic probe interacts with the target molecule, it induces a change in light absorption. Through this alteration, the concentration of the target molecule or other chemical properties can be detected. The characteristic of this instrument is that it does not require an absorption cell. The probe can be directly inserted into the sample solution for in situ determination, without being influenced by external light.

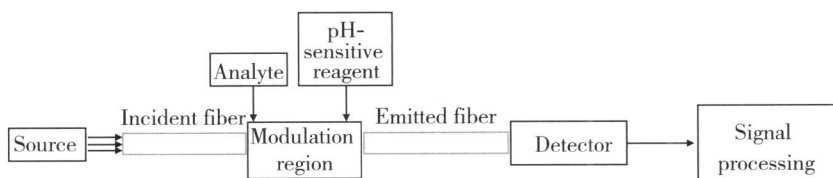

Figure 7. 1　Basic working principle of absorption type fiber optic chemical pH sensor

An absorption type fiber optic chemical pH sensor integrates chemical film formation, optical-fiber technology, and spectrophotometry. The probe is the most crucial component of a pH optical-fiber chemical sensor. It is a reagent-fixing device that can immobilize the sensitive film (whose composition typically consists of pH-sensitive indicators, support agents for fixed reagents, plasticizers, etc.) at the end of the optical fiber or directly coat it on the core of the

　　本实验要求同学们查阅文献，选择合适的材料和方法在光纤探头涂覆一层敏感膜，组装光纤pH传感器，应用于pH值的测定。请拟出实验方案（含实验原理、试剂和仪器、制膜方法、实验条件的选择、pH测定方法和数据处理等），经指导教师审阅后进行实验并写出研究报告。研究报告参照期刊论文的格式书写，包括题目、摘要、关键词、引言、主体部分（实验和结果讨论）、总结、参考文献等。

optical fiber with the cladding corroded, constituting the sensor probe.

In this experiment, students are required to consult the literature, select appropriate materials and methods to coat a layer of sensitive film on the optical-fiber probe, assemble an optical-fiber pH sensor, and apply it to the determination of pH values. Please draft an experimental scheme (including experimental principles, apparatus and chemicals, film-forming methods, selection of experimental conditions, pH determination methods, and data processing, etc.), undergo review by the supervising teacher, conduct the experiment, and write a research report. The research report should be written in the format of a journal paper, including title, abstract, keywords, introduction, main part (experiment and results discussion), summary, and references.

参考文献

[1] 武汉大学. 分析化学实验：上册 [M]. 北京：高等教育出版社，2011.

[2] 李发美. 分析化学实验指导[M]. 北京：人民卫生出版社，2007.

[3] 王艳玮，马兆立. 分析化学实验[M]. 北京：化学工业出版社，2020.

[4] 李红英，全晓塞. 分析化学实验[M]. 北京：化学工业出版社，2018.

[5] 王新宏. 分析化学实验：双语版[M]. 北京：科学出版社，2009.

[6] 蔡明招. 分析化学实验[M]. 北京：化学工业出版社，2004.

[7] 王卫平. 分析化学实验：英汉双语版[M]. 北京：科学出版社，2019.

[8] 于永丽. 分析化学实验[M]. 沈阳：东北大学出版社，2019.

[9] 南京大学大学化学实验教学组. 大学化学实验[M]. 2 版. 北京：高等教育出版社，2010.

[10] 孙玉凤，刘春玲，厉安昕，等. 分析化学实验[M]. 北京：清华大学出版社，2019.

[11] 黄应平. 分析化学实验：英汉双语教材[M]. 武汉：华中师范大学出版社，2012.

[12] 金文英，聂瑾芳. 分析化学实验[M]. 北京：化学工业出版社，2020.

[13] 应敏. 分析化学实验[M]. 杭州：浙江大学出版社，2015.

[14] SKOOG D A，WEST D M，HOLLER F J，et al. Fundamentals of analytical chemistry [M]. 10th ed. Boston，MA：Cengage，2021.

[15] CHRISTIAN G D，DASGUPTA P K，SCHUG K A. Analytical chemistry [M]. Hoboken，NJ：Wiley，2013.

[16] YIN M，GU B，AN Q F，et al. Recent development of fiber-optic chemical sensors and biosensors：mechanisms，materials，micro/nano-fabrications and applications [J]. Coordination chemistry reviews，2018（376）：348–392.

[17] XIE X，DENG Y，PENG J，et al. Nanoparticle-reinforced silica gels with enhanced mechanical properties and excellent pH-sensing performance [J]. Particle & particle systems characterization，2020（37）：1900404.